Palgrave Studies in the History of Science and Technology

Series Editors
Roger D. Launius, Auburn, AL, USA
James Rodger Fleming, Centre Hall, PA, USA

Designed to bridge the gap between the history of science and the history of technology, this series publishes the best new work by promising and accomplished authors in both areas. In particular, it offers historical perspectives on issues of current and ongoing concern, provides international and global perspectives on scientific issues, and encourages productive communication between historians and practicing scientists.

Ivan Malara

Galileo and the *Almagest*, c.1589–1592

How Ptolemaic Astronomy Influenced Galileo's
Early Writings on Motion

Ivan Malara
Department of Philosophy
University of Milan
Milan, Italy

ISSN 2730-972X   ISSN 2730-9738 (electronic)
Palgrave Studies in the History of Science and Technology
ISBN 978-3-031-70613-4   ISBN 978-3-031-70614-1 (eBook)
https://doi.org/10.1007/978-3-031-70614-1

This research was funded by the Department of Philosophy "Piero Martinetti" of the University of Milan under the Project "Departments of Excellence 2023–2027" awarded by the Ministry of University and Research (MUR).

This Palgrave Macmillan imprint is published by the registered company Springer Nature Switzerland AG
The registered company address is: Gewerbestrasse 11, 6330 Cham, Switzerland

... nihil est enim in historia pura et inlustri brevitate dulcius.
(*Cicero*, Brutus, *262*)

To my parents and brother

PREFACE

This book came into existence almost by chance. Its three principal Chapters (2–4) were originally conceived and written as separate articles intended for publication in distinct journals. In their fundamental aspects, they took shape in a span of less than a year, from November 2022 to September 2023. For the sake of convenience, I presented them collectively for the scrutiny of my initial and patient readers. It was their counsel that prompted me to publish all three under a unified title. They urged me to embark on a less commonly traversed path today, leading to the publication of a first monograph not tethered to my doctoral thesis. Despite grappling with numerous uncertainties, I decided to heed their advice. This decision entailed appending a substantive introduction to the subject matter, rectifying errors and oversights in the existing content, expanding the bibliography comprehensively, and concluding with some general observations regarding the scope and limitations of my study. Indeed, this was the course I undertook. Naturally, the final outcome, however it may be evaluated, rests solely on my shoulders.

While I could not draw upon an extensive research project akin to a doctoral thesis, the central figure of this book, Galileo Galilei, has been the focus of my research since my undergraduate years. The University of Milan afforded me the opportunity to continue my studies on Galileo, concentrating on a challenging yet underexplored theme concerning the methodologies employed by the Pisan scientist in the study of one of antiquity's most renowned scientific works: Claudius Ptolemy's *Almagest*.

Fortunately, Ptolemy boasts a rich research tradition and a substantial body of literature. Additionally, a tool accessible to all greatly facilitated my work—the database resulting from the PAL project (*Ptolemaeus Arabus et Latinus*: https://ptolemaeus.badw.de/start), overseen by the Bavarian Academy of Sciences (Bayerische Akademie der Wissenschaften). Without this invaluable resource, locating and studying the sources for my research within such a brief timeframe would have been an insurmountable challenge. The efforts of the team led by Dag Nikolaus Hasse proved indispensable to me in the past year and a half, as essential as the online archive *Galileo//thek@* of the Galileo Museum (https://www.museogalileo.it/it/biblioteca-e-istituto-di-ricerca/progetti/teche/827-galileo-theka.html) has been since the beginning of my studies.

These two databases, in conjunction with various other resources, have provided me with optimal conditions for conducting my research. Nowadays, consulting primary sources is considerably more accessible than it was three decades ago, making it a responsibility to do so. In this book, I have chosen to prioritize the voices of the sources, both in the main text (in translation) and in footnotes (in the original language). The overarching objective of this book is to allow the sources to speak for themselves, enabling them to mutually illuminate each other and thereby restoring a meaning that might otherwise be lost. While this objective is undoubtedly challenging, I would consider myself gratified if I were able to achieve it even in part.

Galileo's *De motu antiquiora*, the source from which the study presented here originated, has been fully translated into English by Raymond Fredette. His translation is readily available online as part of the extensive collection in the ECHO database of the Max Planck Institute for the History of Science (https://echo.mpiwg-berlin.mpg.de/content/scientific_revolution/galileo). In this book, I carried out all translations from *De motu antiquiora*, always keeping a close eye on Fredette's translation. All the other translations from Latin and Italian texts are my own, unless stated otherwise.

Milan, Italy   Ivan Malara

ACKNOWLEDGMENTS

Many people have helped me with my research, making this book possible. I thank Eileen Reeves and A. Mark Smith for their careful reading of early drafts. I am also grateful to Riccardo Bellé, Sven Dupré, Jacqueline Feke, David Juste, Salvatore Ricciardo, and Matteo Valleriani for their assistance and encouragement. I deeply appreciate the support from the Philosophy Library at the University of Milan and the Library of the Galileo Museum in Florence.

Thanks to the participants of the "Peurbachian" seminar, organized by Richard L. Kremer and Razieh S. Mousavi, for our enriching meetings. Special thanks to the CeSIM (*Centro per la Storia Intellettuale del Medioevo*) group in Milan, and especially to Luigi Campi, for fostering informal yet thought-provoking discussions about our works in progress. Consulting with colleagues like David Del Bianco, Michele Meroni, and Luigi Valletta has greatly contributed to the improvement of this book.

Finally, I extend my gratitude to Luca Bianchi, Michele Camerota, Franco Giudice, Niccolò Guicciardini, and Maarten Van Dyck for their guidance. Above all, I thank Elio Nenci for his unwavering support and mentorship.

Also, I am grateful in advance to everyone who approaches this book with a critical mindset. I am indeed deeply convinced that history exists within a boundary that historians continuously corroborate and redefine through well-founded arguments and critiques. This form of constructivism requires the collaboration of multiple trained minds and serves as the most effective means to avoid both historical denialism and mythologies rooted in political ideologies or personal beliefs.

Contents

About the Author

Ivan Malara is a post-doctoral fellow at the University of Milan, Italy. His current research focuses on Galileo Galilei's reception of Ptolemy, with a general interest in exploring the intricate tapestry of sources that may have influenced Galileo's work.

Abbreviations

BB   Bernardino Baldi, *De le vite de matematici* (trascrizione), vol. II, Boncompagni 156 (cat. 1862), 65 (cat. 1892), Albani 619 (?). Available online: https://echo.mpiwg-berlin.mpg.de/ECHOdocuView?url=/mpiwg/online/permanent/library/470S7R08/pageimg&start=1&pn=1&mode=imagepath.

EN   Galileo Galilei, *Opere*. National Edition, ed. by Antonio Favaro, 20 vols. Firenze: Barbèra, 1890–1909.

ENA   Galileo Galilei, *Opere*. Appendix to the National Edition, ed. by M. Camerota et al., 4 vols. Firenze: Giunti Editore, 2013–2019.

GBT   Giovanni Battista Teofilo's Latin translation of Theon's commentary on the *Almagest*. Paris, Bibliothèque nationale de France, lat. 7263. Available online: https://ptolemaeus.badw.de/ms/113/175/1r.

LIST OF FIGURES

CHAPTER 1

Introduction

Abstract The first chapter addresses the challenge of studying Galileo's reception of Ptolemy's *Almagest* due to limited sources. It suggests starting with Galileo's early work, *De motu antiquiora*, which references Ptolemy and mentions a commentary by Galileo on the *Almagest*. This chapter emphasizes the need for careful methodology to avoid speculative theories, such as those proposed by Stillman Drake, and provides an overview of both the *Almagest* and *De motu antiquiora*.

Keywords Galileo Galilei · Claudius Ptolemy · Nicolaus Copernicus · *Almagest* · *De motu antiquiora* · Stillman Drake · Contextualization

Galileo's Copernican beliefs were grounded in Ptolemy's *Almagest*. This may seem like a paradoxical statement, but it is Galileo himself who clarifies its meaning in 1615, together with his pupil and friend Benedetto Castelli. The two, in response to some criticisms raised by Ludovico Delle Colombe, maintain that in order to comprehend the Copernican system, it is necessary to be acquainted with the mathematics of the *Almagest*.[1]

---

[1] "Or se volete ricevere un buon consiglio, desiderando voi [i.e., Delle Colombe] d'intendere 'l Copernico per potergli contradire, mettetevi a studiar prima gli Elementi di

© The Author(s), under exclusive license to Springer Nature
Switzerland AG 2024
I. Malara, *Galileo and the* Almagest, *c.1589–1592*, Palgrave Studies in
the History of Science and Technology,
https://doi.org/10.1007/978-3-031-70614-1_1

Instead of being paradoxical, the opening line now seems common-place. After all, today it is generally recognized that "Copernicus came to his discovery, not by observing the planets more closely, but by understanding Ptolemy more deeply than any of his predecessors. [...] Copernicus was one of the last, and one of the most accomplished, of Ptolemaic astronomers."[2] Galileo, too, was well aware of this. In the Third Day of the *Dialogue on the Two Chief World Systems, Ptolemaic and Copernican* (1632), he argued through Salviati that, initially, "Copernicus restored astronomy upon the assumptions of Ptolemy." Then—Salviati continues—Copernicus realized that he had to change assumptions if he wanted to understand, as an "astronomer–philosopher," the true order of the world. However, a few lines later, Salviati metaphorically maintains that Copernicus healed Ptolemy, stating that "the maladies are in Ptolemy, and their remedies are in Copernicus."[3]

But in 1615, just a year before the Sacred Congregation of the Index issued its anti-Copernican decree, Galileo and Castelli also stated that once the mathematics required to understand *De revolutionibus* is learned, it is impossible not to embrace the Copernican doctrine. So, even Delle Colombe, after acknowledging the mathematical principles common to both Ptolemy and Copernicus, would then

> change [his] mind about Copernicus, and will ascertain how impossible it is to understand him and not to agree with his opinion.[4]

Euclide, cominciando dalla definizion del punto; proccurate poi d'intendere la Sfera e le Teoriche; e intese queste, passate all'Almagesto di Tolommeo, e usate ogni studio per impossessarvene bene; e guadagnata questa cognizione, applicatevi al libro delle Revoluzioni del Copernico" (EN, IV, p. 589, ll. 4–9). For an English paraphrase, see Drake 1999, pp. 290–291 n. 25; English translation by Sergio Knipe in Malara 2023, p. 471.

[2] Evans 1998, p. 425.

[3] See EN, VII, p. 369, ll. 7–17, including the margin text: "Copernico restaurò l'astronomia sopra le supposizioni di Tolomeo"; and ll. 31–32: "Sono in Tolomeo le infermità, e nel Copernico i medicamenti loro." On ll. 7–17, Galileo seems to allude to the fact that Copernicus, in *De revolutionibus*, writes about realizing the falsity of Ptolemaic assumptions only after adopting them. However, Copernicus is not so explicit (see Galilei 1998, vol. II, pp. 712–713).

[4] "[...] e succedendovi il far acquisto di questa scienza, verrete prima a chiarirvi che la cognizion delle matematiche non è da fanciulli, mentre l'andate misurando con quella parte che ne possedete voi adesso; ma misurandola con quello che ne seppe Tolommeo e 'l Copernico, e che allora ne intenderete voi ancora, la troverete essere studio da uomini di

No choice exists as to whether to accept or reject the Copernican system, just as it is impossible to deny that two and two make four. The complaints of a Dostoevskian character or the impositions of a *1984*-like Party would hold little weight. In other words, a strong Ptolemaic education in astronomy imposes an obligatory choice on those who read *De revolutionibus*. Obviously, for Galileo and Castelli, freedom would have been the opportunity to spell this out, to say that two plus two make four—as George Orwell put it. Sadly, still eighty-five years after Galileo's condemnation and abjuration (22 June 1633), their opinion, that it is impossible to disagree with the Copernican doctrine once it is understood, was expunged from the 1718 Florence edition of Galileo's works. It is noteworthy that in 1744, a new edition including the *Dialogue* was printed in Padua and welcomed with the approval of Pope Benedict XIV. However, the opinion expunged in 1718 was not restored in 1744.[5]

At this point, the opening sentence appears strong, extreme, even exaggerated.[6] In fact, it is precisely insofar as a hyperbole that it serves to convey at least a twofold message. The first is that mathematics is not a childish discipline, easy to learn, but rather "a discipline for men aged a hundred," worthy of the highest speculative minds.[7] The fact that it is potentially accessible to everyone, provided they invest much effort and study, does not diminish but rather elevates its status. The second message, somewhat more specific, is that the truth of the Copernican system is evident only in the light of the *Almagest*.

This second message is valuable to us, readers centuries later, because it is doubly indicative. On the one hand, it suggests that Galileo approached *De revolutionibus* more as a mathematician educated in customary astronomy teachings than as a philosopher dissatisfied by Aristotle's physics and cosmology—although the two aspects did run

cent'anni; e, quello che vi sarà più maraviglioso, cangerete opinione intorno al Copernico, e vi accerterete come è impossibile l'intenderlo e non concorrer con la sua opinione" (EN, IV, p. 589, ll. 9–16). Knipe's translation (see *supra*, n. 1) slightly modified.

[5] See Galilei 1718, vol. I, p. 472, and Galilei 1744, vol. I, p. 437. On the 1744 edition, see Finocchiaro 2005, pp. 126–155.

[6] Galileo and Castelli deliberately placed excessive emphasis on the role of mathematics in the Copernican conversion. Both were well aware that, even for the very few who possessed an outstanding understanding of Ptolemaic astronomy, this was usually not enough to change their minds after reading *De revolutionibus*.

[7] See *supra*, n. 4. Valuable insights into the sixteenth-century debates on mathematics can still be found in De Pace 1993.

together in Galileo's work. On the other hand, it enlightens us about Galileo's fundamental approach to the text of *De revolutionibus*, a factor that undoubtedly played a crucial role in his conviction as a Copernican. Apparently, only a reader deeply knowledgeable in astronomy, to the extent of also knowing the mathematical technicalities of the *Almagest*, could have been able to understand and, therefore, agree with Copernicus' arguments. Not only the arguments accessible to everyone, presented in the "first sketch" (*prima dipintura*) of *De revolutionibus*— namely the first book—but primarily those expounded in the subsequent five books, which require a good mathematical preparation.[8]

Is it possible that all of this was more than a rhetorical counteroffensive against Ludovico Delle Colombe? In order to attempt an answer, one should first address another question, which has been hitherto overlooked: what knowledge did Galileo have of Ptolemy, and specifically of the *Almagest*? Considering the breadth of the question, this book aims to provide a groundwork and initiate further inquiry on the subject.

Research and Methodology

To begin with, it is crucial to inquire whether undertaking this research is worthwhile. Can we delve into Galileo's reception of the *Almagest*? At first glance, it appears unlikely. Remarkably, no comprehensive studies of this nature have been undertaken thus far. This raises suspicion, particularly given the extensive Galilean bibliography comprising over 25,000 entries![9] Consequently, at least two possibilities emerge: either there is a lack of interest in the subject at hand, or it is impractical to explore.

In the first scenario, rekindling interest could be achieved through targeted research to illuminate aspects that have been, perhaps inadvertently, overlooked. In the second scenario, however, there is little

---

[8] "[...] voi [Delle Colombe] non avete, non dirò intese le sue demonstrazioni, ma né capite le semplici ipotesi, né anco i nudi termini dell'arte, né intesa la *prima dipintura* che mette il Copernico nel principio del suo libro" (EN, IV, p. 588, emphasis added). See also EN, XI, *Galileo to Gallanzone Gallanzoni* (16 July 1611), p. 153: "[...] il nostro S. Colombe non ha pur vedute *le 2 prime et più facili carte ad essere intese*, dove il Copernico per sua principalissima hipotesi pone che la sfera stellata sia altissima di tutte e totalmente immobile" (emphasis added).

[9] See the census provided by the Museo Galileo: https://www.museogalileo.it/it/biblioteca-e-istituto-di-ricerca/progetti/banche-dati-e-bibliografie/508-bibliografia-galileiana.html (last visited on 28 December 2023).

recourse. It implies insufficient testimony, evidence, or documentary sources to work profitably. This constitutes the worst-case scenario for the historian. In this case, the demarcation line between the novelist and the historian becomes so subtle that it may risk fading away. For the historian, this risk occurs indeed when sources, generally understood as the media of information on history, are scarce and vague, allowing for conjectures.

Despite not having the intention to write a novel, I believe that conjectures should not be avoided on principle. On the contrary, it is the responsibility of the historian to develop them properly. While crafting this concise book, I deemed well-grounded those conjectures that rely solely on sources or other types of evidence immediately available for scrutiny by other expert scholars. To put it plainly, a conjecture is well-grounded when it is based on more than one source, and the entirety of its sources is checkable. This definition should be understood as my personal benchmark for rigor in writing this book, rather than a rigid, abstract rule excluding all other forms of historical conjectures.

A pertinent example might help clarify this seemingly complex discourse. The second chapter of this book revolves around an autograph testimony from Galileo regarding his hope for the imminent publication of his commentary on the *Almagest*.[10] This testimony, dating perhaps to the period of his teaching in Pisa (1589–1592), is the sole surviving piece of evidence for such a commentary. The commentary was never published and has not been found among Galileo's unpublished manuscripts that have come down to us.

Based on this testimony, one could hypothesize that Galileo's commentary on the *Almagest* did indeed exist. Beyond its existence, hypotheses about its contents could be formulated with primarily circumstantial evidence drawn from other works. Then, by combining selectively chosen statements from Galileo, similar to the one presented at the beginning of this introduction, one could even conjecture that the lost commentary on the *Almagest*, thus the *Almagest* itself, had a significant role in Galileo's scientific journey.

Thus, Galileo's testimony, circumstantial evidence, and ad hoc statements from Galileo could create the impression for the reader of facing a well-grounded, well-argued conjecture about the past. However, this conjecture—that is, the relevance of the *Almagest* to the development of

[10] See *infra*, Chapter 2, n. 1.

Galileo's science—is not entirely checkable, as it relies on an unavailable source, namely Galileo's commentary on the *Almagest*.[11] It is a legitimate conjecture, of course. However, to assess its plausibility, one is compelled to shift to the methodological side, noting, for example, that Galileo's testimony, in the absence of other corroborating testimonies or conclusive documents, does not constitute historical proof of the existence of his alleged commentary on the *Almagest*. Consequently, it is easy to discern that the entire attempt to reconstruct its contents and make a conjecture about its importance is nothing more than a beautiful yet fragile sandcastle.

To resist the (rather strong) temptation of constructing magnificent sandcastles or embarking on Pindaric flights, I have chosen to limit myself as much as possible to formulating well-grounded conjectures. That is, as stated above, conjectures that can be immediately checked, as they are based on sources accessible to everyone. Therefore, in the second chapter, instead of speculating about Galileo's alleged commentary on the *Almagest*, I considered whether there was a necessity, at that time, to compose such a commentary, even to the extent of pretending to have already written it. To attempt to answer this question, I adopted the classic method of contextualization. In this case as well, however, I have consciously chosen to restrict my work to a form of contextualization understood as the examination and comparison of contemporary texts, likely known to Galileo.

It is certainly useful to interpret the history of science in the light of great economic, social, political, and geopolitical contexts.[12] I believe,

---

[11] Without venturing too far, a similar observation can be made regarding a conjecture that is frequently accepted, specifically concerning the significance of Jesuit science in the formation of Galileo's scientific ideas. This claim is grounded in the contents of Mss. Gal. 27 and 46, now transcribed, respectively, in EN, I, pp. 15–171, and ENA, III, pp. 21–99 (on these manuscripts, see Carugo & Crombie 1983, and Wallace 1984b). However, as of today, the actual Jesuit sources from which the two manuscripts were copied have yet to be uncovered (see Malara 2019). Moreover, the notion of "Jesuit science" is too vague, and upon closer inspection, the connection between Mss. Gal. 27, 46, and the subsequent development of Galileo's scientific thought is not apparent (see Dollo 2003, pp. 87–128).

[12] Within the realm of Galilean studies, a noteworthy exploration of this kind has been successfully conducted regarding Galileo's telescope (see Bucciantini & Camerota & Giudice 2015).

however, that it is equally beneficial to do so by focusing on a micro-history, as it were, derived from a small selection of intellectual productions. Common aspirations, contrasting ideas, specific idiosyncrasies, and so forth, can indeed emerge from an array of carefully selected texts. In this way, it becomes possible to outline a framework within which individual statements acquire their contextual meaning and historical sense. I hope, in this manner, to have demonstrated in the second chapter that when Galileo alluded to his alleged commentary on the *Almagest*, he was attempting to intercept a common need shared by many scholars of Ptolemy's masterpiece.

Although even a single testimony, when placed in context, can be profitably examined, it is undeniably true that the almost complete absence of studies on Galileo's reading of the *Almagest* largely arises from an apparent scarcity of available material. Delving into the relationship between Galileo and the *Almagest*, therefore, carries a significant risk of leading to highly speculative conjectures—conjectures that do not contribute in any way to illuminating our past. Certainly, they contribute to generating interest in the chosen theme. However, while aiming to bring attention to a less-explored topic within the extensive literature on Galileo, this goal cannot and should not justify the means. Conversely, it is the means at the historian's disposal—namely, a valid methodology—that justify a certain result, albeit one that is provisional and subject to improvement.

On this note, it is essential to remember that the connections between Galileo and Ptolemy's work have actually undergone examination, at least on one occasion. However, the results were rather problematic, as the intended goal seemed to have influenced the means.

Stillman Drake, a distinguished scholar of Galileo, addressed this subject toward the end of the 1970s. More precisely, his essay titled *Ptolemy, Galileo, and Scientific Method* was published in 1978, presenting a thought-provoking exploration. This essay was subsequently reissued in 1999 as part of Drake's collection of articles titled *Essays on Galileo and the History and Philosophy of Science*.[13] Drake's exploration originated from a well-known proposition: Galileo's scientific method lacked foundational philosophical ideas or convictions. According to Drake, philosophical subtleties were not the bedrock but rather a consequence of Galileo's

---

[13] I will refer exclusively to the 1999 re-edition of this text.

science.[14] By asserting this, Drake intentionally distanced himself from the viewpoints of other respected scholars, including Alexandre Koyré. The latter, for instance, positioned Plato's philosophy as the cornerstone of Galileo's scientific thought.[15] In Drake's perspective, method and science were synonymous for Galileo. Science constituted a method, that is, a series of procedures that, in its execution, did not necessitate presupposing any philosophical ideas about nature.[16]

Irrespective of the ambiguity in the terms used in this debate about the origins of Galileo's science (what kind of Platonic philosophy did Galileo adhere to?[17] Can there be a method lacking theoretical foundations? What exactly is meant by philosophy when stating that Galileo's science was devoid of it?), it is crucial to observe that Drake asserted a specific thesis. He posited that Galileo had adopted and refined a method proposed around the second century CE by Claudius Ptolemy at the outset of his *Almagest*.[18]

This method entailed the use of mathematics, recognized as the sole discipline truly capable of producing certain knowledge, applied to the study of celestial phenomena to predict their behavior. In practice, mathematical hypotheses, when tested against everyday experiences related to celestial bodies, would evolve into practical theoretical models essential for compiling astronomical tables. Ultimately, if these tables proved effective, they could only affirm the hypotheses upon which they were grounded.[19]

This procedural approach does not delve into the causes of the predicted phenomena and often relies on abstract models stemming

---

[14] See Drake 1999, p. 274. For a fresher and well-defined take on the matter, see Hatfield 1990, and Van Dyck 2022a.

[15] See Drake 1999, pp. 288–289 n. 1.

[16] Without commenting on Drake's thesis, I will note that he supported it by referencing a passage from his translation of *The Assayer* where the term "method" is mentioned (see ibid., p. 275). However, this term does not appear in Galileo's original text (see EN, IV, p. 237, ll. 10–16).

[17] On this matter, see Galluzzi 1973.

[18] See Drake 1999, esp. pp. 275–284. However, Drake specified that the *Almagest* was the "principal source" of Galileo's view on science and method, not the only one (see ibid., pp. 275, 282).

[19] See ibid., p. 278.

from idealization.[20] According to Drake, this is precisely the method that Galileo clearly acknowledged and extended beyond the realm of astronomy into the realm of physical inquiry:

> The new science of motion developed by Galileo mainly during 1602-1609 and presented to the public in 1638 was worked out along lines entirely compatible with the 1602 statement that so closely paralleled Ptolemy's program, simply extended from astronomy to physics.[21]

The reference made by Drake in this passage to the statement from 1602 pertains to the beginning of Galileo's *Trattato della sfera ovvero cosmografia*, a work published posthumously that explains some fundamental principles of geocentric astronomy.[22] Drake acknowledged the observation made by Willy Hartner, emphasizing the derivative nature of Galileo's treatise on cosmography, which, consequently, lacks any innovative elements compared to traditional teachings on the matter.[23] However, Drake posited that the opening passage was absent in similar contemporary texts and was later added by Galileo around 1602.[24]

Specifically, he claimed that in the initial sentence, Galileo contends that Ptolemy's method, expounded shortly afterward without explicit mention of Ptolemy, can be applied to all sciences, including physics. I present Drake's translation:

---

[20] See ibid., pp. 280–281. To delve into the significance of idealization in Galileo's work, see Van Dyck 2018, which also offers bibliographical details on the subject.

[21] Drake 1999, p. 280.

[22] See EN, II, pp. 203–255.

[23] See Drake 1999, p. 275, where Drake quotes Hartner 1967, p. 184. Yet there are some interesting features in Galileo's treatise on the sphere. For instance, Galileo wrongly ascribes to Ptolemy the argument about the stone thrown from a tower (see EN, II, p. 224), which was most likely derived from Brahe's *Epistolae astronomicae* (see Camerota 2004, p. 102). As for the sources of Galileo's treatise on cosmography, see Cardoso & De Andrade Martins 2008, Cardoso & De Andrade Martins 2017, and De Andrade Martins 2010.

[24] See Drake 1999, pp. 276, 284–288. The argument was also presented in Drake 1978, pp. 51–55.

In the Treatise on the Sphere, which we may more properly call Cosmography, as in the other sciences, the subject should be identified and then the order and method to be observed in this should be touched on.[25]

Actually, it is evident that in this passage Galileo is advocating a rather traditional thesis.[26] When delving into the explanation of any scientific discipline, it is initially essential to define the specific object of study for that discipline and then elaborate on the approach to be taken. Subsequently, Galileo asserts that when engaging in cosmography, the studied object is the world or the universe (which I will also take as synonyms throughout this book). Therefore, if Drake's analysis, stating that the method elucidated here by Galileo is common to all properly designated sciences, is accurate, it implies that all sciences focus on studying the world. Which is a bizarre statement, to say the least. Furthermore, Galileo distinctly delineates which aspects of the world fall under the purview of the cosmographer on the one hand, and which belong to the natural philosopher on the other hand.[27]

However, Drake did not appear to be troubled by this particular aspect, as he was confident that there were numerous other pieces of evidence

---

[25] Drake 1999, p. 276. A slightly different translation was provided in Drake 1978, p. 52. The original reads as follows: "Nel Trattato della Sfera, che più propriamente chiameremo Cosmografia, prima, sì come in tutte l'altre scienze, si deve avvertire il suo suggetto, ed in oltre toccare qualche cosa dell'ordine e metodo da osservarsi in esso" (EN, I, p. 211).

[26] See *Met.* K (XI), 7, 1063 b 36–1064 b 14 (already in *Met.* E (VI), 1, 1025 b 3–1026 a 32). In 1969, Fredette placed this passage from Aristotle's *Metaphysics* next to a passage from Buonamici's *De motu* (see Fredette 1969, pp. 60–61, 327 n. 69).

[27] "Diciamo dunque, il suggetto della cosmografia essere il mondo, o vogliamo dire l'universo, come dalla voce stessa, che altro non importa che *descrizione del mondo*, ci viene disegnato. Avvertendo, però, che delle cose, che intorno ad esso mondo possono esser considerate, una parte solamente pertiene al cosmografo; e questa è la speculazione intorno al numero e distribuzione delle parti d'esso mondo, intorno alla figura, grandezza e distanza d'esse, e, più che nel resto, intorno a i moti loro; lasciando la considerazione della sostanza e delle qualità medesime al filosofo naturale" (EN, II, p. 211). English translation in Drake 1999, p. 276.

supporting his thesis.[28] To demonstrate that Galileo's method lacks philosophical origins, he selectively cited passages from Galileo and juxtaposed them with Ptolemy's *Almagest*. This effort to substantiate his initial thesis bears little relevance to the historian of science.

In the first place, the *Almagest* is treated as a book devoid of history. Drake relied on an English translation (occasionally modified) by Catesby Taliaferro from 1952, based on the Heiberg edition of 1907.[29] Does it make sense to compare this translation of the *Almagest* with Galileo's texts? From a historical point of view, it does not. Drake remained silent on the nature of Galileo's knowledge of the *Almagest*. Which version of the *Almagest* did he read? Or more precisely, what edition or editions could he actually consult? How was the *Almagest* circulated during Galileo's time?

Even the author of the *Almagest*, Ptolemy, seems like a figure detached from time and history in Drake's article. Is it truly evident that Ptolemy crafted a non-philosophical method for astronomy? Was he genuinely not considered a philosopher in his time?[30] And even if he was not, how was he perceived in Galileo's time? What image or images of Ptolemy were circulating back then?

While Drake's article presents intriguing ideas, it falls short of establishing a solid groundwork for an in-depth exploration of Galileo's interest in the *Almagest*. Not only does it leave lingering questions unanswered, but it also fails to bring them into consideration.

To avoid the pitfalls of Drake's article, I have chosen to apply the contextualization method from the initial chapter to both the third and fourth chapters. The extensive footnotes do not serve as a showcase of erudition, which unfortunately I lack, but align with the checkability

---

[28] "Whether Galileo did mean that what he said here was to be applied elsewhere is best decided by what he did in other sciences, later, recalling Einstein's advice" (ibid., p. 277). Einstein's advice runs as follows: "If you wish to learn from the theoretical physicist anything about the methods he uses, I would give you the following advice: Don't listen to his words, examine his achievements. For to the discoverer in that field, the constructions of his imagination appear so necessary that he is apt to treat them not as the creations of his thoughts but as given realities" (ibid., p. 275).

[29] See ibid., p. 289 n. 12.

[30] Today, we appreciate Ptolemy also as a philosopher, although he does not neatly fit into a monolithic framework, such as the strictly Aristotelian or Platonic one. See Feke 2018.

criterion elucidated earlier. Within these footnotes, the discerning historian can discover exhaustive or nearly complete citations of numerous primary sources underpinning my study. The reader will then be able to check whether my interpretations and translations are plausible, forced, or totally incorrect.

However, many queries remain unanswered. For example, it is difficult to determine whether Galileo consulted just one edition of the *Almagest* (and if so, which one), or more than one.[31] In such instances, I have consistently endeavored to be maximally thorough, citing from the three main Latin editions theoretically accessible to Galileo.[32] Another example revolves around the inquiry into how Ptolemy was perceived in Galileo's time. Addressing this question necessitates a dedicated study, which, to my knowledge, remains unexplored. In the third chapter, I just noted that, conceivably, individuals of that period might have adhered to Ptolemaic views in astronomy without necessarily subscribing to Aristotelian principles in natural philosophy.

Additionally, due to the constraints of limited time, I have focused my scrutiny on a single work. This particular work, known as *De motu antiquiora*, is pivotal today for comprehending, at least in part, the development of Galileo's scientific and philosophical thought.

THE *ALMAGEST* AND GALILEO'S *DE MOTU ANTIQUIORA*

The *Almagest* requires no extensive introductions. Undoubtedly, it stands as one of the most renowned scientific works in history, perhaps comparable in significance to Isaac Newton's *Philosophiae naturalis principia mathematica* (1687). Penned by the Alexandrian Claudius Ptolemy around the mid-second century CE, the *Almagest* has significantly shaped the study of astronomy for over a millennium.[33]

As succinctly captured by David Juste:

---

[31] See *infra*, the beginning of Chapter 3.

[32] I excluded the 1538 Greek edition (see *infra*, Chapter 2, n. 84). I am not aware of any vernacular rendition of the *Almagest* circulating in the sixteenth century.

[33] According to Olaf Pedersen, the *Almagest* "was just as important to ancient science as Newton's *Principia* was to the seventeenth century, and there is no question that it was a greater scientific achievement than the *De revolutionibus* which has obliterated its fame, just as Copernicus has outshone Ptolemy as astronomical genius" (Pedersen 2011, p. 11).

The *Almagest* [...] is generally considered the culmination of Greek astronomy and the single most important astronomical work until Copernicus's *De revolutionibus orbium coelestium* (1543). It provides comprehensive mathematical models explaining all celestial movements in a geocentric universe and offers all the necessary numerical data to calculate the positions of the seven planets for any time, past, present and future. The *Almagest* comprises 13 books dealing respectively with cosmology and trigonometry (Book I), spherical astronomy (II), the Sun (III), the Moon (IV-V), eclipses (VI), the fixed stars (VII-VIII), Mercury (IX), Venus and Mars (X), Jupiter and Saturn (XI), retrogradation (XII) and planetary latitudes (XIII).[34]

Prior to influencing European culture, the work received meticulous attention from Arab astronomers. The current title, *Almagest*, directly stems from Arabic roots. The original Greek title was *Mathēmatikē syntaxis* (*Mathematical composition*), also referred to as *hē megalē syntaxis* (*The great composition*), or simply *megiste* (*The greatest*). The superlative was then calqued into Arabic as *al-majisti*, and later Latinized into *Almagesti* or *Almagestum*. Even from its title alone, one can glean the cross-cultural importance that the work held.[35]

The most important Latin translation of the *Almagest* can be traced back to the twelfth century, accomplished by Gerard of Cremona, who rendered it in Arabic.[36] This translation made the inherent complexity of the *Almagest* even more apparent, necessitating efforts to enhance the accessibility of Ptolemy's teachings, particularly in Europe, where astronomy found its place in numerous universities as part of the *quadrivium*.

---

[34] Juste [19].

[35] As James Evans has suggestively written, "the history of the Western astronomical tradition, with its Greek, Arabic, and Latin contributions, is embodied in this book title" (Evans 2018, p. 790). On the Latin title *Almagesti*, see Ptolemy 1515, f. 1r: "[Ptolemaeus] fecit libros multos, de quorum numero iste est, qui Megasiti dicitur, cuius significatio est Maior perfectus, quem ad linguam volentes convertere Arabicam nominaverunt Almagesti." See also Kunitzsch 1974, pp. 15–71.

[36] This translation was made in Toledo before 1175 and subsequently revised until 1187, coinciding with the death of Gerard of Cremona. It was crafted based on two Arabic versions and exists in two distinct forms, with one being a revision of the other. On this matter, the main reference remains Kunitzsch 1974. See also Kunitzsch 2008, Burnett 2010, esp. pp. 1–2, and Burnett 2013. Additionally, see Juste [3], and the list of references cited therein.

Among the works dedicated to such an endeavor, it is worth recalling at least four: (1) the *Almagesti minor* (c. 1200), or *Almagestum parvum*, which presented the first six books of the *Almagest* in a Euclidean fashion[37]; (2) the *Sphere* (early thirteenth century) by Sacrobosco (John of Holywood), achieving significant success as an introductory work to the first book of the *Almagest*—it drew also from another important textbook on the *Almagest*, namely al-Farghani's *Rudimenta astronomica*[38]; (3) the *Theoricae novae planetarum* (1454) by Georg (von) Peurbach, which rapidly supplanted various preceding attempts to elucidate the mathematical models employed by Ptolemy to expound the apparent motions of each planet (including the Sun and the Moon)[39]; (4) the *Epitoma Almagesti*, a detailed summary of the entire *Almagest*, initiated by Peurbach shortly before his demise (April 1461) and concluded by his pupil Johannes Regiomontanus (Johannes Müller von Königsberg).[40]

It was during the latter half of the fifteenth century that the *Almagest* started to attract renewed and noteworthy interest, particularly owing to humanist scholars like the aforementioned Peurbach and Regiomontanus, as well as figures such as Basilios Bessarion and George of Trebizond. Bessarion revived interest in the original words of Ptolemy's *Almagest* in Latin Europe, while George of Trebizond produced a new Latin translation of it from Greek. This translation, and many other works related to the *Almagest*, were printed (or reprinted) throughout the sixteenth century.[41]

This century also witnessed the birth of Galileo Galilei in 1564 in Pisa. His early education, from primary studies to university courses, occurred in the last decades of the sixteenth century. Furthermore, his career as a university professor of mathematics commenced in 1589 in Pisa. In 1592,

---

[37] See Zepeda 2015, Zepeda 2018.

[38] Thorndike 1949 remains an important reference work on the *Sphere*. Within the long tradition of the *Sphere*, the groundbreaking research led by Matteo Valleriani has resulted in numerous publications. The most relevant for this study are Valleriani 2020 and Valleriani & Ottone 2022. On al-Farghani's *Rudimenta astronomica*, or *Elements of astronomy*, see Abdukhalimov 1999.

[39] See Aiton 1987, and Malpangotto 2021.

[40] See Zinner 1990, pp. 51–55, and n. 51 on pp. 213–214. See also Malpangotto 2008, pp. 33–36.

[41] For the conflict that emerged around the *Almagest* between George of Trebizond and Bessarion, see Shank 2017, pp. 50–65. See also *infra*, Chapter 2. On the first printed editions of the *Almagest*, see the accurate overview in Pellacani 2020, pp. 47–54.

he moved to Padua, where he taught mathematics for eighteen years. Then, in 1610, with the publication of the *Sidereus Nuncius*, everything changed for Galileo.[42]

The *Sidereus Nuncius* was the third publication bearing Galileo's name, yet it can definitively be considered his first groundbreaking work.[43] In this book, at the age of forty-six, he publicly confessed his Copernican convictions for the first time.[44] Apparently, Galileo's telescopic observations convinced him that the time had finally come to expose publicly the deficiencies of Aristotelian natural philosophy and geocentric cosmology. However, based on the information available, it appears that sometime before 1597, he had already recognized the superiority of Copernicus over Aristotle and Ptolemy. Despite this realization, he refrained from making his views public, fearing that the majority of people would not understand him.[45]

[42] For a comprehensive study on Galileo's life and work, see Camerota 2004, undoubtedly the most thorough bibliography on Galileo.

[43] The first to be published was the *Operazioni del compasso geometrico militare* in 1606 (see EN, II, pp. 363–424). On this instrument, see Favaro 1966, vol. I, pp. 165–192, Drake 1977, Valleriani 2010, pp. 27–41. The second publication, released a year later in 1607, was a response to Baldassarre Capra, who plagiarized the *Operazioni* (see EN, II, pp. 513–599).

[44] This can be observed from the outset in Galileo's dedication to Cosimo II, where it is stated that the four "stars" of Jupiter, along with Jupiter itself, revolve around the Sun: "quae [i.e. quatuor Sidera] quidem disparibus inter se motibus circum Iovis Stellam caeterarum nobilissimam, tanquam germana eius progenies, cursus suos orbesque conficiunt celeritate mirabili, interea dum unanimi concordia circa mundi centrum, circa Solem nempe ipsum, omnia simul duodecimo quoque anno convolutiones absolvunt" (EN, III, p. 56). As noted by Camerota, "le scoperte esposte nel *Sidereus Nuncius* rivestivano, almeno agli occhi del loro autore, un marcato significato filo-copernicano" (Camerota 2004, p. 169). On Galileo's Copernicanism in the *Sidereus Nuncius*, see especially Giudice 2014, pp. 53–60.

[45] See EN, X, *Galileo to Kepler* (4 August 1597), pp. 67–68, esp. 68: "in Copernici sententiam multis abhinc annis ven[i], ac ex tali positione multorum etiam naturalium effectuum caussae s[u]nt a me adinventae, quae dubio procul per commune hypothesim inepxplicabiles sunt. Multas conscripsi et rationes et argumentorum in contrarium eversiones, quas tamen in lucem hucusque proferre non sum ausus, fortuna ipsius Copernici, praeceptoris nostri, perterritus, qui, licet sibi apud aliquos immortalem famam paraverit, apud infinitos tamen (tantus enim est stultorum numerus) ridendus et explodendus prodiit. Auderem profecto meas cogitationes promere, si plures, qualis tu es, extarent: at cum non sint, huiusmodi negotio supersedebo." This letter has been thoroughly examined and contextualized by Massimo Bucciantini (see Bucciantini 2003, pp. 49–68).

So, it seems that it is still the Cinquecento, as it approaches its end, the century in which some important views of Galileo took shape, in natural philosophy as well as in cosmology and astronomy. Fortunately, a series of survived manuscripts provides an interesting glimpse into Galileo's intellectual endeavors roughly between 1589 and 1592, when he was teaching at the University of Pisa. Since they are unified by the treatment of motion, the more mature Galileo probably gathered them together under the title *De motu antiquiora scripta mea*, which translates to *My earlier writings on motion*. Today, they are usually referred to as *De motu antiquiora*, or simply *De motu*.[46]

Before even finding a place in the National Edition of Galileo's works curated by Antonio Favaro, *De motu antiquiora* had already captured the attention of numerous Galileo scholars. The reasons for this interest are diverse. First and foremost, it is crucial to note that within Ms. Gal. 71, which houses the majority of *De motu antiquiora* documents, there exists a fragment of *De motu naturaliter accelerato* from a decidedly later period. However, initial scholars failed to discern the posterior nature of the fragment, attributing to the early Galileo ideas about free fall motion that were developed later, akin to those published in the Third Day of the *Two New Sciences* (1638). In truth, a meticulous examination of *De motu antiquiora* reveals a pronounced contrast between the theory of free fall acceleration advocated therein and that found in the fragment of *De*

---

[46] The title *De motu antiquiora* is drawn from Viviani and Nelli (see EN, I, *De motu* (*Avvertimento*), pp. 245–246). However, it is not found in the current manuscript layout. In the national edition, Favaro opted for the title *De motu*. Fredette suggested that the original title most likely was *De motu antiquiora scripta mea* (see Fredette 1969, p. 370, n. 172, and Fredette 1972, p. 327). Regarding dating, see Fredette 1969, pp. 139–163, summarized in Fredette 1972, pp. 331–333. Fredette's compelling analysis, based on internal and external evidence, provides a high likelihood of dating the work between 1589 and 1592. Additionally, see Camerota 1992, pp. 69–101, where arguments against the alternative dating proposed in Carugo & Crombie 1983 are presented. Alongside the already mentioned works of Fredette and Camerota, see also Drabkin 1960, Fredette 2001, and Fredette 2017. For a detailed overview of the Pisan context in which *De motu antiquiora* took shape, see Camerota & Helbing 2000.

motu naturaliter accelerato. The contrast lies in the very conception of acceleration.[47]

The more mature Galileo, perhaps the one more familiar to us, considered free fall acceleration as a form of uniformly accelerated motion. In this type of motion, starting from rest, an object gains velocity proportional to the time elapsed since the initial moment of descent.[48] Additionally, this mature perspective led Galileo to withhold definitive conclusions regarding the causes of this phenomenon. The acceleration of free fall, in a sense, was attributed to a natural inclination to motion inherent in all bodies. However, the precise nature of this inclination—also known as gravity—and its underlying cause remained unclear to him.[49]

---

[47] See Camerota 1992, pp. 17–38, 103–139. Fredette made the following distinction among scholars (see Fredette 1969, pp. 5–31): some saw *De motu antiquiora* as almost entirely successful ("un succès quasi total") due to the inclusion of the fragment *De motu naturaliter accelerato*, while others viewed it as nearly a complete failure ("un échec quasi total") when the fragment was excluded. Fredette also noted that, although Favaro maintained that *De motu antiquiora* cointained "in germe" the later developments of Galileo's science, he was cautious enough not to insist on this (see EN, I, *De motu* (*Avvertimento*), p. 246).

[48] Galileo did not reach this conclusion right away. Despite discovering the quadratic proportionality that linked the spaces traversed and the time elapsed in the free fall of an object, he initially associated this "accident" with an "axiom" stating that velocity increases in direct proportion to the spaces traversed. This is evident in Galileo's famous letter to Sarpi dated October 16, 1604 (see EN, X, p. 115). For an analysis of the "spontaneity" of thinking in terms of space traversed rather than time elapsed, see Koyré 1966. Camerota provides compelling evidence that Galileo realized and corrected his initial mistake between 1607 and 1609 (see Camerota 2004, pp. 144–148). For an in-depth examination of the early evolution of Galileo's science of motion, based on Ms. Gal. 72, see Büttner 2019. Van Dyck 2022b can be read as a more accessible introduction to Büttner's book.

[49] On gravity as an inherent inclination common to all heavy bodies, see EN, VIII, *Discorsi e dimostrazioni*, p. 118, ll. 27–31, where Salviati states that "un corpo grave ha da natura intrinseco principio di muoversi verso 'l commun centro de i gravi, cioè del nostro globo terrestre, con movimento continuamente accelerato, ed accelerato sempre egualmente, cioè che in tempi eguali si fanno aggiunte eguali di nuovi momenti e gradi di velocità." And some pages later, he says: "Non mi par tempo opportuno d'entrare al presente nell'investigazione della causa dell'accelerazione del moto naturale, intorno alla quale da varii filosofi varie sentenzie sono state prodotte [...]; le quali fantasie, con altre appresso, converrebbe andare esaminando e con poco guadagno risolvendo" (ibid., p. 202, ll. 19–27). According to Paolo Galluzzi, Galileo made every effort to uncover the cause of free fall acceleration, thereby giving a mechanical foundation to his science. Yet, ultimately, he had to relinquish the pursuit (see Galluzzi 1979, pp. 311–329, 384).

In *De motu antiquiora*, however, the early Galileo viewed free fall acceleration as an external and passing phenomenon. He also identified its cause in the momentary presence of an *impetus* or *virtus impressa*, which bestowed a kind of lightness to the falling object. Gradually diminishing, this *virtus* allowed the falling body to progressively regain its own weight. Once regained, the body would then continue to fall at a uniform velocity. This velocity, in turn, was not considered directly proportional to the weight of the falling body and inversely proportional to the resistance of the medium, as argued by Aristotle. Instead, Galileo, drawing inspiration from Archimedes' hydrostatics, asserted that the velocity of fall should be proportional to the difference between the weight of the falling body and that of an identical volume of the medium in which it falls, such as air, for instance. This perspective also facilitated the departure from the Aristotelian concept of absolute lightness. The upward rectilinear motion of fire could, in fact, be attributed to what we might anachronistically term "hydrostatic thrust" today. This is because the weight of a certain volume of fire is less than that of the same volume of air.[50]

It is evident, therefore, that a substantial disparity exists between Galileo's thoughts in *De motu antiquiora* and the fragment *De motu naturaliter accelerato*. Among Favaro's many merits, one must undoubtedly acknowledge his decision to exclude this fragment from *De motu antiquiora*.[51] For the sake of clarity, in this book, when referring to *De motu antiquiora*, I am specifically addressing the series of six writings grouped as follows in the first volume of the Edizione Nazionale (EN)[52]:

[50] On the notion of *virtus impressa* used by Galileo, see Fredette 1972, pp. 338–342. For a general overview on the notion of *impetus* in the Renaissance, see Van Dyck & Malara 2022. To delve into Galileo's so-called "Pisan dynamics," it remains valuable to study Clavelin 1996, pp. 130–148, and Galluzzi 1979, pp. 166–197. Concerning hydrostatic thrust, it was explicated at that time through the concept of "extrusion," drawing upon the doctrines of ancient atomists. In *De motu antiquiora*, Galileo refrains from promptly embracing this notion. Its inclusion only becomes evident in subsequent revisions. See Fredette 1969, pp. 270–277, and Galluzzi 1979, pp. 175–176 n. 89.

[51] The fragment was published by Favaro in EN, II, pp. 261–266. See also his *Avvertimento* on pp. 259–260.

[52] See EN, I, *De motu* (*Avvertimento*), pp. 247–248. Camerota included Ms. Gal. 71, f. 60v (blank) in those containing the treatise in 10 chapters (see Camerota 1992, p. 61). Due to certain internal evidence in the text, Fredette discerned a division into two books within the treatise (see Fredette 1969, p. 38, and Fredette 1972, pp. 334–335). However, in this study, I do not find it necessary to adopt such a division.

1. Treatise in 23 chapters	EN, I, pp. 251–340 (= Ms. Gal. 71, ff. 61r-124v)	
2. Treatise in 2 chapters	EN, I, pp. 341–343 (= Ms. Gal. 71, ff. 133r-134v)	
3. Treatise in 10 chapters	EN, I, pp. 344–366 (= Ms. Gal. 71, ff. 43r-60r)	
4. Dialogue	EN, I, pp. 367–408 (= Ms. Gal. 71, ff. 4r-35v), of which pp. 375, l. 10–378, l. 3 (= Ms. Gal. 46, ff. 102r-104v)	
5. Memoranda	EN, I, pp. 409–417 (= Ms. Gal. 46, ff. 102r, 104v-110r)	
6. Plan of work	EN, I, pp. 418–419 (= Ms. Gal. 71, f. 3v)	

As evident, there exist two primary compositions: one in the format of a treatise and the other in a dialogic structure. Concerning the treatise, at least three versions can be discerned (1–3). In contrast, the dialogic form has left us with a sole rendition (4), depicting a conversation between two characters, Alexander and Dominicus. Additionally, there are preliminary notes, categorized into memoranda (5) and a plan of work (6).

Another aspect that has captivated scholars' attention regarding *De motu antiquiora* pertains to their relative chronology. According to Favaro, as illustrated in the aforementioned diagram, all versions of the treatise precede the dialogue. Many scholars have critiqued this arrangement based on internal evidence. Presently, after extensive studies, a nearly unanimous consensus among scholars seems to have been reached. Galileo initially composed the dialogue between Alexander and Dominicus, subsequently crafting the treatise in 23 chapters. Following this, he revisited the treatise, reworking its initial two chapters. Finally, he undertook a subsequent revision and rewriting of the initial ten chapters. I find this relative chronology to be persuasive. Consequently, in this book, I have assumed the dialogue was written first and have denoted the three renditions of the treatise simply as the first version (23 chapters), the second version (2 chapters), and the third version (10 chapters).[53]

The partial insights gathered thus far offer a glimpse into why these writings have consistently fascinated Galileo scholars. Their stratification and the individuation of the "Pisan dynamics"—a doctrine of motion strongly influenced by Archimedes, openly rejecting Aristotelian principles, though still intertwined with traditional ideas like that of *impetus*—are just a couple of reasons that highlight the importance of Galileo's

---

[53] Giusti 1998 offers the most comprehensive analysis of the relative chronology of *De motu antiquira*, also presenting readers with a wealth of references related to this historiographical debate.

earlier writings on motion. In essence, *De motu antiquiora* provide a window into the philosophical and scientific musings of Galileo before he assumed the well-known persona we are acquainted with, even if only through myths, legends, and popular narratives.

To grasp the intellectual trajectory of Galileo, it seems fitting to scrutinize *De motu antiquiora* in light of a comprehensive understanding of the subsequent developments in his thinking. However, such an examination carries the risk of inadvertently succumbing to a form of anachronism. For instance, it may involve placing *De motu antiquiora* within a linear, almost teleological evolutionary framework of Galileo's thought, as if his early studies on motion were an indispensable prerequisite for the later ones. This assumption is anachronistic as a linear progression of his thought—devoid of deviations, fractures, or drastic changes—may not necessarily align with the reality of the historical facts.[54] It might be our subsequent projection.

A more tangible example pertinent to the theme of this book concerns Galileo's Copernicanism. In *De motu antiquiora*, Galileo explicitly adheres to a geocentric cosmological model.[55] Nevertheless, certain assertions regarding the possibility of a perpetual circular motion of a sphere around the center of the world, as well as about the motion of a sphere along a plane parallel to the horizon, may be read as implying a Copernican position. Over a century ago, Emil Wohlwill was the first to propose that, in *De motu antiquiora*, Galileo was already a supporter of the Copernican model.[56]

The passages considered by Wohlwill to substantiate his thesis are primarily three, presented below in the chronological order that Wohlwill

---

[54] It is noteworthy that Galileo repurposed some parts of *De motu antiquiora* to draft the *Discorso delle cose che stanno in sull'acqua*, released in 1612, a period during which he had already relinquished numerous Pisan theories. For a comprehensive examination of this, see Camerota 2000, esp. the comparison on pp. 81–82.

[55] Many are the passages where he says that heavy bodies move toward the center of the world. See, for instance, *infra*, Chapter 3, n. 24.

[56] See Wohlwill 1884, esp. pp. 74–82. This old paper is now accessible online: https://www.digi-hub.de/viewer/image/DE-11-001661160/82/LOG_0015/ (last visited on 15 January 2024). Other scholars, such as Wolff, Koyré, and Naylor, concurred that some assertions in *De motu antiquiora* exhibit a Copernican influence. See bibliographical references in Büttner 2008, p. 38 n. 11.

believed they were written (he indeed considered Favaro's proposed relative chronology of *De motu antiquiora* to be accurate)[57]:

A

A mobile that has no external resistance naturally descends on a plane, however little inclined below the horizon, with no external force applied, as is evident in water. And the same mobile, when placed on a plane, however little inclined above the horizon, ascends only when a force is applied [*non nisi violenter*]; therefore, it remains that on the horizon itself, it neither moves naturally nor violently. If it does not move violently, then it can be moved by the smallest force of all.[58]

B

Therefore, natural motion occurs when movable objects, in their motion, approach their proper place; on the contrary, violent motion occurs when movable objects, set in motion, move away from their proper place. Since things are this way, it is evident that a sphere revolving around the center of the world is not moved by either natural or violent motion. For, since the sphere is heavy, and the center of heavy bodies is the center [of the world], and heavy bodies move according to [the motion of] their own center of gravity, if the center of gravity of the sphere were already at the center of the world, where it would [locally] rest while the sphere revolves around it, then it is evident that the sphere would not be moved either naturally or violently, as it would neither approach nor recede from its proper place. [...] If the sphere were at the center of the world, whether it would be moved perpetually or not must be considered, assuming the motion is imparted by an external mover. [...] This is not the place for an answer; for it must first be determined from what those things that are not moved naturally are set in motion.[59]

---

[57] I refer here to Wohlwill 1909, pp. 105–110, which is a later assessment by Wolhwill more centered on Galileo. Wohlwill 1884 indeed had a more general focus on inertia. Furthermore, in 1909 Wolhwill could refer to Favaro's National Edition, whereas in 1884, he had to reference Alberi's so-called "Granducale" edition.

[58] "Mobile, nullam extrinsecam habens resistentiam, in plano sub horizonte quantu-lumcunque inclinato naturaliter descendet, nulla adhibita vi extrinseca; ut patet in aqua: et idem mobile in plano quantulumcunque super horizontem erecto non nisi violenter ascendit: ergo restat, quod in ipso horizonte nec naturaliter nec violenter moveatur. Quod si non violenter movetur, ergo a vi omnium minima moveri poterit" (EN, I, p. 299). On the so-called principle of minimal force, see Festa & Roux 2006.

[59] "Motus itaque naturalis est dum mobilia, incedendo, ad loca propria accedunt; violentus vero est dum mobilia, quae moventur, a proprio loco recedunt. Haec cum ita se habeant, manifestum est, sphaeram super mundi centrum circumvolutam neque naturali

C

If a marble sphere were at the center of the world, so that the center of the world and the center of the sphere were the same, and if, then, the beginning of the motion of the sphere were given by an external mover, then the sphere perhaps would not move by violent motion but by a natural one. In fact, there would be no resistance at the axis, and the parts of the sphere would neither approach nor recede from the center of the world. However, I said "perhaps" because if such motion were not violent, it would last forever. Yet, this eternity of motion seems far from the nature of the earth itself, for which rest seems more pleasant [*iucundior*] than motion.[60]

None of these quotations explicitly references the Copernican doctrine. Nevertheless, according to Wohlwill, it is possible to discern between the lines ("zwischen den Zeilen") of Galileo's words an implicit allusion to

---

neque violento motu moveri. Cum enim sphaera gravis sit, et gravium locus sit centrum, moveanturque gravia secundum suae gravitatis centrum; si iam sphaerae esset centrum gravitatis in centro mundi, in quo, dum sphaera circumducitur, maneret; manifestum est quod neque naturaliter nec violenter moveretur, cum ad proprium locum nec accederet nec recederet. [...] Si sphaera esset in centro mundi, nec naturaliter nec violenter circum-ageretur, quaeritur, utrum, accepto motus ab externo motore, perpetuo moveretur nec ne. [...] Non est hic responsionis locus; videndum enim prius est, a quo moveantur quae non naturaliter moventur" (EN, I, pp. 304–306). In order to understand this passage, one should also read the following memorandum: "Motum localem appellamus illum, in quo mobilis centrum gravitatis movetur: quare caelestium orbium motus locales non dicemus, cum eorum centrum gravitatis, quod magnitudinis etiam centrum est, immobile semper maneat" (EN, I, p. 416). From this, one might infer that, according to the early Galileo, celestial orbs exist, are heavy, and lack local motion. This will be briefly discussed in the last chapter devoted to concluding remarks.

[60] "Si itaque marmorea sphaera existeret in centro mundi, ita ut centrum mundi et centrum sphaerae essent idem, deinde initium motus sphaerae a motore externo daretur, tunc sphaera fortasse non moveretur motu violentu sed naturali; cum nulla ibi esset axium resistentia, nec partes sphaerae centro mundi accederent aut recederent. Dixi autem, fort-asse: quia si talis motus non esset violentus, perpetuo duraret; ista autem motus aeternitas ab ipsius terrae natura longe abesse videtur, cui quies iucundior quam motus esse videtur" (EN, I, p. 373).

the Earth's double circular motion.[61] It may be, however, that this interpretative approach is highly influenced by our knowledge of Galileo's later thoughts. Especially in hindsight, *De motu antiquiora* can be seen as Galileo's initial, yet tentative and rudimentary attempt at justifying the Copernican doctrine. I am not asserting that such an interpretation is incorrect, but neither can we rule out the possibility that it might be a perspective illusion.[62] Even Wohlwill himself, in the end, exercised great caution when interpreting those passages as unmistakably Copernican.[63]

Consider, for instance, that both Copernicus and the mature Galileo viewed the Earth's double circular motion as natural.[64] However, in text A and B, Galileo delves into a motion that defies classification as either natural or violent.[65] Toward the end of B, he argues that understanding the perpetuity of this motion hinges on comprehending what generates

[61] "Von dieser [i.e., the Copernican theory] ist freilich in den Pisaner Heften nirgends ausdrücklich die Rede. Das einzige Zitat aus dem Buche des Copernicus das sich in ihnen findet, nimmt nur auf eine geometrische Erörterung Bezug [i.e., EN, I, p. 326], deren weitere Verwertung für die Zwecke des Astronomen nicht berührt wird. Um so bedeutsamer erscheint, was die Untersuchung über die Bewegung im Kreise *zwischen den Zeilen* verrät" (Wohlwill 1909, pp. 107–108, emphasis added).

[62] According to Jochen Büttner, for example, the quoted passages should not be read "as theoretical speculations situated in a cosmological context." On the contrary, he argued that "Galileo's argument was more plausibly located in the context of the questions that arose for the early modern engineer-scientists from contemporary machine technology, in particular from the increasing application of so-called flywheels" (Büttner 2008, p. 34).

[63] While in 1884 he claimed that in *De motu antiquiora* there is "unzweideutiges Zeugnis [unambiguous evidence] für Galileis Beschäftigung mit der Copernicanischen Lehre in der Pisaner Periode" (Wohlwill 1884, p. 82, n. 2), in *Galilei und sein Kampf für die copernicanische Lehre* his interpretation became more nuanced: "Als Zeichen einer frühen Beschäftigung mit der copernicanischen Lehre wurden die besprochenen Bestandteile der Pisaner Aufzeichnungen zu deuten versucht; dagegen gestatten eben diese Ausführungen keinen einwandsfreien Schluß auf Galileis Gesinnung in dem gleichen Zeitpunkt. Obgleich seine Erwägungen der Erdbewegung günstig scheinen, sind sie doch nach Form und Inhalt nicht minder mit der Denkweise des vorsichtig zweifelnden Forschers als mit der des überzeugten Anhängers vereinbar; sie können aus dem Bestreben hervorgegangen sein, aber auch aus dem Verlangen nach Widerlegung selbstempfundener Bedenken" (Wolwhill 1909, p. 109).

[64] See ibid., p. 108, where he quotes *De rev.* I.8: "Verum si quispiam volvi terram opinetur, dicet utique motum esse naturalem, non violentum." On this, see Copernicus 2015, vol. III, p. 103, n. 3. The notion that circular motion is natural is clearly articulated by Galileo throughout the First Day of the *Dialogue*.

[65] On this, see Wolff 1987, pp. 243–246, and Camerota 1991, pp. 194–195. Additional information in Helbing 1989, pp. 229–232.

it—whether its cause is perpetual or self-consuming. Yet this issue is never addressed in *De motu antiquiora*. Additionally, the center of the world is explicitly identified as the proper place for all heavy bodies, including the Earth. Only a forced and somewhat biased connection between text A and the Earth's revolutionary motion of the heliocentric system might lead one to believe that, when writing *De motu antiquiora*, Galileo already adhered to Copernican views. Text C, on the other hand, is highly ambiguous. It could be interpreted in a diametrically opposing manner, as has indeed been done, compared to Wohlwill's proposal.[66] It should also be noted that, in that passage, Galileo does not present a third option beyond natural and violent motion. Wohlwill probably interpreted this *tertium non datur* as a step forward, compared to A and B, toward the Copernican idea of the Earth's natural circular motion.[67] However, it is uncertain whether text C was penned after A and B. In fact, as previously noted, most scholars today compellingly believe that C (i.e., the dialogue) was composed before A and B (i.e., the first version of the treatise).

It is important to note that I am not dismissing the possibility that, during that period, Galileo was genuinely a committed Copernican. I am confining myself to a methodological observation on the most suitable approach for a historian to adopt when interpreting *De motu antiquiora*. I am of the opinion that the most effective approach does not involve making general assumptions about whether Galileo adhered to the Copernican model or not.

Let us make another pertinent example. There is a passage in *De motu antiquiora* where Galileo refutes Aristotle's theory of *quies media*. According to this theory, rectilinear natural motion must pause before reversing direction.[68] Galileo lists some arguments against this theory, including the following:

---

[66] For instance, Calvelin, referring to the same passage, concluded that "l'ontocosmologie [aristotélicienne], toujours présente, bloque immédiatement toute possibilité d'ébranler sérieusement la dichotomie des mouvements naturels et des mouvements violents" (Clavelin 2016, p. 77).

[67] He considered text C as an "einer zweiten, ohne Zweifel späteren Bearbeitung der gleichen Frage" (Wohlwill 1909, p. 106).

[68] *Phys.* VIII, 262 a 12–263 a 2. This theory is also recalled by Simplicio in the Third Day of the *Dialogue* (see EN, VII, p. 301, ll. 17–19). On this, see also Galilei 1998, vol. II, p. 629.

The third argument can be derived from a certain rectilinear motion, which Copernicus constructs from two circular motions in his *De revolutionibus*. There are indeed two circles, one of which is carried on the circumference of the other. While the carried circle moves faster than the other, a point on it moves in a straight line and continuously inverts direction [*regreditur*] through the same; nevertheless, it cannot be said to be at rest at the extremes, since it is continuously carried around by the circumference of the circle.[69]

This is a clear reference to Copernicus' version of the so-called "Tusi couple" exposed in *De rev.* III.4.[70] This shows Galileo had reached at least this chapter of *De revolutionibus* when he wrote the first version of the *De motu antiquiora* treatise.

However, Anna De Pace has recently argued that at that time Galileo "perhaps had not even read *De revolutionibus*."[71] She believes that Galileo might have obtained similar information not directly from Copernicus, but from Giambattista Benedetti's *Diversarum speculationum liber* (1585), or even from Paolo Sarpi, whom Galileo met in Padua. In so doing, De Pace takes up some hints already given by Michele Camerota, but, contrary to the latter, she uses them to support the hypothesis that Galileo was not familiar with *De revolutionibus*.[72]

[69] "[Terti]um argumentum desumi potest a motu quodam recto, quem ex duobus circularibus motibus Nicolaus Copernicus in suis Revolutionibus componit. Sunt enim duo circuli, quorum alter in alterius circumferentia fertur, cuius signum unum, dum alter altero citius movetur, in recta fertur linea et per eandem continue regreditur; nec tamen dici potest, illud in extremis quiescere, cum continue a circuli circumferentia circumducatur" (EN, I, p. 326).

[70] See Camerota 2004, p. 100; Copernicus 2015, vol. II, pp. 177–179, and vol. III, pp. 256–257.

[71] "Il quale quadro concettuale [i.e., that of *De motu antiquiora*] lascia pensare che quando Galileo scriveva i *De motu* forse non aveva nemmeno letto il *De revolutionibus*, o comunque certamente non aveva aderito alle sue dottrine" (De Pace 2020, p. 532; in the footnote on p. 533 it is stated that Galileo mentions Copernicus also in the dialogued version of *De motu antiquiora*, namely in EN, I, p. 386. However, this is not the case).

[72] See Camerota 2004, p. 100: "Di certo – è un fatto sicuramente desumibile dall'esame dei primi lavori galileiani – negli anni giovanili Galileo ebbe modo di leggere il *De revolutionibus orbium coelestium* di Copernico"; "la menzione fattane nei *De motu* attesta [...] in modo piuttosto chiaro, un attento studio dell'opera dell'astronomo polacco"; "il suo studio del capolavoro copernicano dovette essere oltremodo approfondito e consapevole." For the references to Benedetti and Sarpi, see ibid., p. 587, n. 97.

However, it is important to emphasize that Benedetti's text, also recalled by Jacopo Mazzoni in his *De comparatione*, does not imply the double circle mechanism used by Copernicus.[73] Secondly, if Sarpi truly was Galileo's source, this might push the timeline of *De motu antiquiora* forward, aligning it with Galileo's intensive study of Copernicus—the opposite of what De Pace aims for. Lastly, the fact highlighted by De Pace that the Tusi couple had been known for a long time and circulated through the works of Albert of Brudzewo, Regiomontanus, Amico, Fracastoro, and others, does not prove how Galileo could, based on any of these works, attribute such a mechanism to Copernicus.[74]

In general, according to De Pace, Galileo fully embraced Copernicanism upon grasping "Copernicus's theory of gravity, rooted in that of Plato and Plutarch."[75] However, since Galileo, in *De motu antiquiora*, asserts that all heavy bodies tend toward the center of the world, De Pace infers that at that time he might not have been well-acquainted with

---

[73] See Mazzoni 1597, p. 193s; Mazzoni 2010, pp. 297–298. It seems that, as noted by Mazzoni, Benedetti drew inspiration from Regiomontanus and Peurbach: "Tertius error eiusdem philosophi [i.e., Aristotelis] ob contemptum mathematicarum est quia credidit duos motus rectos, quorum alter esset sursum, alter vero deorsum, vel dextrorsum et sinistrorsum, vel ante et retro, vel (ut compendio dicam) eundo et redeundo per eandem lineam non posse esse continuos, sed necessario inter illos quietem aliquam interponi. Hoc enim profitetur se demonstrasse in octavo libro Physicorum non difficili et satis longa ratiocinatione, contra quam instant mathematici, demonstrando contrarii possibilitatem. Assumatur enim *figura*, in qua indicant Astrologi lineam directionis et retrogradationis planetarum, *quam nobis exhibuerunt Ioannes Regiomontanus 4 libro compendii et Georgius Purbachius in principio secundae partis libelli Theoricae Planetarum*" (ibid., emphasis added). The "figura" reproduced in Mazzoni's book is drawn from Benedetti's *Diversarum speculationum liber*. As for Mazzoni's reference to Peurbach's *Theoricae novae*, he perhaps referred to Reinhold's commentary, where the "Secunda pars libelli" concerns the "Planetarum passiones," and presents at the outset a figure titled "Schema progressuum stationum et regressuum." I am not sure which figure from Regiomontanus' *Epitoma*, book IV, Mazzoni had in mind.

[74] "Si può aggiungere che il meccanismo che combina il moto uniforme di due cerchi per produrre un moto rettilineo (la cosiddetta coppia di Tūsī) era noto da tempo e di esso si erano già avvalsi Fracastoro e Amico, e ancor prima Alberto di Brudzewo e Regiomontano" (De Pace 2020, p. 533, n. 31).

[75] "È dopo aver acquisito la teoria della gravità di Copernico, radicata in quella di Platone e Plutarco, che Galileo nella Prima Giornata del *Dialogo* avrebbe argomentato che, se la Terra fosse grave, occorrerebbe pensare a un altro intero verso il cui centro il globo terrestre, 'essendone rimosso, cercasse di ritornare' [...]" (De Pace 2020, p. 533). On this, see also her interesting observations ibid., pp. XXIII–XXXVI.

Copernicus' theory of gravity. Thus, he had not yet embraced Copernicanism. This argument is consistent, but it cannot contribute to our understanding of *De motu antiquiora* in any meaningful way. In fact, it may lead us to exclude a more nuanced scenario where Galileo's shift from being non-Copernican to becoming a convinced Copernican is not as linear and clear-cut as one may simplistically believe. This may indeed be the scenario depicted in *De motu antiquiora*.[76]

What is actually evident from *De motu antiquiora* is that Galileo has at least a partial understanding of *De revolutionibus* but never explicitly endorses Copernicus' theories. On the contrary, he explicitly supports a geocentric cosmology, despite his strongly anti-Aristotelian tone. It becomes intriguing, therefore, to delve into the characteristics of the geocentrism articulated by Galileo in these earlier writings on motion, irrespective of whether he genuinely embraced it.

By adopting this approach, I believe it is possible to avoid many pitfalls while attempting to comprehend the role of Ptolemy and the *Almagest* in Galileo's *De motu antiquiora*. Indeed, another undeniable piece of evidence from *De motu antiquiora* is Galileo's references to specific chapters of the *Almagest*. This book is primarily dedicated to examining and elucidating these references.

GROUNDWORK

In summary, this book presents an initial, concise exploration of Galileo's study of Ptolemy's *Almagest*, a topic that has not been thoroughly investigated until now. While Galileo is commonly associated with Copernicus and the abjuration, this work sheds light on a lesser-known phase in his intellectual journey. There was a time when he, at least in a series of writings on motion, appeared to hold geocentrism while vehemently criticizing Aristotle. Galileo's early engagement with geocentrism and Ptolemy can be studied through these early writings, known as *De motu antiquiora*.

Two broad questions have guided my work: How did Galileo and his contemporaries read and study Ptolemy's *Almagest*? Can we gain new insights into Galileo's heliocentrism by examining his early reception of the *Almagest*? These questions have directed my focus, in the second

---

[76] See Camerota 2004, p. 101.

chapter, toward Theon of Alexandria's commentary on the *Almagest*, which was still considered a significant source in the sixteenth and seventeenth-century reception of Ptolemy. Perhaps, it was also a notable source for Galileo, as I have suggested in the third chapter of this book. The same questions have also brought to light, in the third chapter, the possibility of the existence of a Ptolemaic geocentrism distinctly separate, and at times intentionally detached, from Aristotelianism. This is certainly not sufficient to warrant using label of *Ptolemaism* or even its plural form, *Ptolemaisms*. However, it seemed appropriate to emphasize the high level of awareness with which some sixteenth-century authors distinguished the Ptolemaic elements from the Aristotelian ones in the so-called Aristotelian-Ptolemaic system.

All things considered, this book can only offer a preliminary study, laying the groundwork for further explorations on the same topic. Firstly, this is because it focuses solely on one of Galileo's works. To obtain a comprehensive understanding of the relationship between Galileo and the *Almagest*, and how it evolved over time (if it did), a systematic study of the entire *Corpus Galilaeanum* is necessary. Secondly, even though it is based on a single work, the study proposed here on *De motu antiquiora* does not claim to be exhaustive. Nonetheless, and thirdly, the methodological precautions upon which it is built (checkability, contextualization, caution regarding assumptions and conclusions of general significance) could prove beneficial in further investigations into the same theme. In essence, this study aims to open new avenues for inquiry rather than filling an existing gap.

CHAPTER 2

Why a Commentary?

Abstract The second chapter investigates Galileo's claim in *De motu antiquiora* that he had written a commentary on the *Almagest* ready for publication. Given the absence of such a commentary and Galileo's general reluctance to write commentaries, it questions why he mentioned it. The chapter examines writings from contemporary mathematicians and astronomers, like Francesco Barozzi, Christoph Clavius, and Giuseppe Biancani, which reveal a strong demand for a Latin commentary on the *Almagest*. It also discusses Theon of Alexandria's Greek commentary on the *Almagest* and the efforts to translate it into Latin, including Giovanni Battista Teofilo's translation, which was eventually donated to Galileo's pupil, Vincenzo Viviani.

Keywords Galileo Galilei · Claudius Ptolemy · Nicolaus Copernicus · Theon of Alexandria · Sixteenth-century reception of the *Almagest*

Consider the following quotation:

> Yet it is not the water, but the shape of the cup, that causes such an effect [of optical magnification], as we have explained in more detail in our

I. Malara, *Galileo and the* Almagest, *c.1589–1592*, Palgrave Studies in the History of Science and Technology,
https://doi.org/10.1007/978-3-031-70614-1_2

commentary on Ptolemy's *Great Construction* [viz. the *Almagest*], which (God willing) will be published soon.[1]

Neither Galileo's commentary on the *Almagest*, to which this quotation refers, nor Galileo's *De motu antiquiora* (1589–1592), from which this quotation is taken, were ever published. But while there are various manuscript versions, and preparatory notes left from *De motu antiquiora*, nothing at all remains of the alleged commentary. What happened to it? Despite Galileo's confident assertion that it was on the brink of being published, why can we not find any trace of it? Is it possible that it never even existed?

Among Galileo scholars, some asserted that the commentary existed, while others argued it was simply an unrealized idea or project. Upon consulting the university records (*rotuli*) of the Studio of Pisa, Charles B. Schmitt noted that during his third and final year of teaching mathematics in Pisa (1591–1592), Galileo taught the first book of the *Elements* and general "hypotheses of celestial motions" (*celestium motuum hypotheses*). Michele Camerota cautiously suggested that this course likely served as the basis for Galileo's commentary on the *Almagest*, assuming the latter actually existed.[2]

On a more speculative note, Stillman Drake proposed that Galileo devoted several years to his commentary on the *Almagest*, envisioning it as a significant cosmological work that would bring considerable renown. Drake reconstructed that the commentary initially presented a geo-heliocentric system akin to Tycho Brahe's. However, around mid-1591, Galileo supposedly refined this into a model more similar to Ursus', featuring a diurnally rotating Earth. According to Drake, Galileo developed both systems in Pisa, before he had any knowledge of either Brahe's or Ursus' work or the controversy between them. Only upon his arrival in Padua (end-1592), immersed in a cultural environment more open to scientific innovation, did Galileo become acquainted with the works

---

[1] "Verum non aqua, sed calicis figura, talis effectus causa ut fusius in commentariis super Magnam Ptolemaei Constructionem declaravimus, quae (Deo favente) brevi eduntur" (EN, I, 314). In this text and others, I have translated the plural "*commentarii*" from Latin to the singular "commentary" in English. This is because in Latin, the term usually refers to a series of comments on a text, akin to what we now recognize as a commentary.

[2] See Schmitt 1972, p. 262; Camerota 2004, p. 57.

of Brahe and Ursus. This revelation, Drake conjectured, led Galileo to abandon his plans to publish the commentary on the *Almagest*.[3]

Finally, it is worth mentioning the opinion of William A. Wallace, who argued that Galileo did not write a commentary at all. Instead, he intended to craft a kind of summary of the *Almagest*, drawing heavily from Christoph Clavius' commentary on the *Sphere*. In fact, the matter of objects appearing larger when submerged in a water vessel—a topic Galileo touches on in the opening quote—was also a subject of discussion for Clavius.[4]

As this brief overview shows, the hypotheses proposed so far have been more or less conjectural. It is crucial to clarify from the outset that, to this day, definitive answers remain elusive. We lack sufficient historical documents and testimonies to confirm or refute whether Galileo authored a commentary on the *Almagest*. However, as is often the case, the level of guesswork can be lessened by carefully contextualizing the subject matter.

Yet this kind of study has not been undertaken to date.[5] This has left a significant historiographical gap concerning the influence of Ptolemaic astronomy on Galileo's thought. Indeed, the question of whether Galileo truly wrote a commentary on the *Almagest* implies a series of more profound queries about Galileo's understanding and interpretation of the *Almagest*. What knowledge did Galileo have of the *Almagest*? How did he read and interpret it? Did his understanding of the *Almagest* influence his reading of Copernicus' *De revolutionibus*? If so, how? Did it ease or impede his acceptance of the Copernican system? These critical questions have, up until now, been largely overlooked.[6]

---

[3] See Drake 1987, pp. 100–103.

[4] See Wallace 1981, p. 228. Here Wallace also claims that Clavius' commentary on the *Sphere* "epitomizes the *Almagest* and so could well be the source Galileo had in mind for his project summary." This is an overstatement. At most, one can say that Clavius touches upon some themes from the *Almagest* in his commentary on the *Sphere*. In fact, he intended to write his own commentary on the *Almagest*, as I will discuss later in this chapter.

[5] This is not surprisingly, given the apparent scarcity of historical evidence on the matter. On this matter, see Chapter 1. It suffices here to notice that, to date, only extremely speculative studies have been done, such as the aforementioned Drake 1987, to which Drake 1999 can also be added. The latter is based on a doubtful interpretation of a passage from Galileo's *Trattato della sfera overo Cosmografia*.

[6] For example, these questions are not addressed in De Pace 2020, the most recent study on Galileo's interpretation of Copernicus.

There's another, simpler question that has been bypassed and seems to be an appropriate starting point: why a *commentary* on the *Almagest*?

THE NEED FOR A COMMENTARY ON THE *ALMAGEST*: BAROZZI AND CLAVIUS

It is unusual to think of Galileo as a commentator on other works. The activity closest to commentary that Galileo undertook was perhaps in writing *The Assayer* (1623), wherein he extended his critical annotations to the *Libra* (1619) by Lotario Sarsi, a pseudonym of Orazio Grassi. In fact, while *The Assayer* engages deeply with Lotario Sarsi's *Libra*, it does not quite fit the traditional genre of commentary; it almost stands as its parody.[7]

At that time, a commentary was primarily understood as a work dedicated to the exposition and clarification of another book, from which one was required to depart in order to produce new knowledge. However, the commentary itself encompassed new insights and perspectives that were distinct from the original source text. Moreover, "a 'pure' understanding of the source text was not the exclusive goal."[8] As Karl Enenkel have emphasized, commentaries "had an enormous impact on education; reading and writing practices; the formation, organization, authorization, and transmission of knowledge; and the reception of the classics." Also, they "shaped university education; various professional activities; professional scholarship; private learning and intellectual entertainment; printing and publishing; religious life; and even segments of life that were seemingly far removed from scholarship and learning, such as warfare, engineering, and agriculture."[9] In short, they represented a genre entirely distinct from the educational materials we create today.[10]

---

[7] Recently, Michele Camerota and Franco Giudice have emphasized that *The Assayer* can be considered as "un commentario critico della *Libra*, che vuole anche essere una parodia della tipica analisi scolastica dei libri di Aristotele" (Galilei 2023, p. xxix). John Heilbron also described *The Assayer* as "a mocking or mockery of a scholastic analysis of an Aristotelian text" (Heilbron 2010, p. 245). See also Battistini 2005, p. 89.

[8] Enenkel& Nellen 2013, p. 3.

[9] Enenkel 2014, pp. 3–4.

[10] For an extensive overview of early modern commentaries, refer to Enenkel& Nellen 2013, which provides valuable bibliographic references. For another extremely useful introduction to the topic, see Most 1999 and Grafton 2010.

For numerous centuries, the commentary served as the central genre in scientific discourse. Galileo was well aware of this, not because he wrote commentaries, but rather because he fiercely criticized this type of bookish culture.[11] One can consider Galileo's famous metaphor of the "book of nature"—the sole worthy philosophical book that, in his view, only mathematicians were equipped to decipher.[12]

However, Galileo did not always use this metaphor in his work. It emerged relatively late, and it was more frequently employed after the publication of the *Sidereus Nuncius* in 1610, at a time when Galileo was engaged in defending himself from his most relentless and traditionalist adversaries.[13] It appears, then, that there was a phase when Galileo deemed it beneficial to commit himself to the creation of a commentary. When he wrote *De motu antiquiora* during the early 1590s, he saw indeed the utility of composing a commentary on the *Almagest*. This is a certain fact that persists even when judgment is withheld on whether Galileo wrote such commentary.

When viewed from a material culture angle, it is noteworthy that Galileo's decision or project to write a full commentary on the *Almagest* is consistent with the editorial gaps pointed out by some of his contemporaries. One example is Francesco Barozzi, who outlined several difficulties related to the study of the *Almagest* in the preface to his *Cosmographia* (1585)—a book owned by Galileo.[14] Among these difficulties,

---

[11] Obviously, Galileo was not the only one at the time to make this sort of criticism. In Grafton's succinct and clear phrasing, "many propagandists of the New Philosophy, most notably Bacon, Galileo, and Descartes, insisted that knowledge of nature was the most important of all studies and that textual study could never yield new or effective knowledge in the sphere of nature" (Grafton 2010, p. 231). See also Enenkel& Nellen 2013, p. 68.

[12] See EN, VI, *Il Saggiatore*, p. 232; Galileo's letter to Fortunio Liceti of January 1641, EN, XVIII, p. 295. See also Galilei 2023, pp. 46–47 nn. 161–163.

[13] According to Mario Biagioli, "the genealogy of the topos [of the book of nature] is directly linked to the contingencies of Galileo's engagement with the theologians in the 1613–5 period" (Biagioli 2003, p. 563; see also Biagioli 2006, pp. 233–234). However, Francesco Barreca has rightly pointed out that the "book of nature" metaphor was already used by Galileo in 1611 (see Barreca 2018, p. 128 n. 46). Note that Galileo also used the metaphor in the dedicatory of the *Dialogue* (1632) to present both Copernicus and Ptolemy as philosophers (see Bianchi 2022, p. 301).

[14] See Favaro 1886, p. 260. According to Roberto De Andrade Martins and Walmir Thomazi Cardoso, Barozzi's work could have been one of the sources of Galileo's *Trattato della sfera ovvero cosmografia* (see Cardoso& De Andrade Martins 2017, p. 143).

the first was linked to the solemnity of Ptolemy ("*propter Autoris grav-itatem*") and the incredible complexity of the subject matter under study. Then, there was the issue of the geometric and arithmetic proofs, which appeared "obscure" to scholars of his time. Lastly, there were the corrupted Greek in the *editio princeps* (1538) and the "poor Latin translation" (*mala latina versio*), though it is unclear whether Barozzi was referring to Gerard of Cremona's translation from Arabic (1515) or George of Trebizond's from Greek (1528).[15] Collectively, these challenges made the study of the *Almagest* particularly daunting, especially for "novices" (*Tirunculi*).[16]

However, Barozzi also believed that, despite its flaws, the *editio princeps* provided an essential tool for understanding the *Almagest*, namely, Theon of Alexandria's commentary.[17] This commentary on the *Almagest* was incomplete, but in the *princeps* it was supplemented with Pappus'

---

[15] Cf. Ptolemy 1515, Ptolemy 1528, and Ptolemy & Theon 1538. Barozzi was not the only one to complain about the quality of translations. Regiomontanus, and likely Bessarion too, claimed that if Ptolemy were to come back to life in their time, he would not even recognize his own work because of the "barbaric" renditions of it (see Peurbach& Regiomontanus 1543, p. 2; on the "if Ptolemy were alive / had known" *topos*, see the Del Soldato 2020, pp. 215–216 nn. 65, 66, and Malara 2023, p. 470). Gerard of Cremona's translation was also considered imprecise by Bernardino Baldi, a contemporary of Galileo, who described it as "very imperfect" in his unpublished *Life of Ptolemy* (see BB, f. 340v: "[...] delle traduttioni di questa opera [i.e., the *Almagest*] due s'hanno per le mani, l'una dall'Arabico assai imperfecta [i.e., Gerard of Cremona's], e l'altra dal Greco, e questa è fatica di Giorgio Trapezunzio"). Many years later, in 1673, Marchetti complained about the translation of "Trapesunzio," describing it as "most barbarous and obscure" (*barbarissima e oscurissima*) (Pellacani 2020, p. 53).

[16] "Quod propter Autoris gravitatem, et rerum Astrologicarum maximam difficultatem, demostrationumque Arithmeticarum, et Geometricarum (praesertim hisce nostris temporibus) obscuritatem, ac etiam Graeci exemplaris depravationem, et malam Latinam versionem, non ab omnibus, et praecipue a Tirunculis intelligi facile potest" (Barozzi 1585, f. b5v). Even experts in Euclid's *Elements* found Ptolemy's demonstrations challenging, a fact also emphasized by Jacopo Mazzoni: "[...] liber ille [viz. magna compositio] ob difficillimarum demonstrationum acumen etiam illis qui Euclidis elementa triverunt impervius esse videtur" (Mazzoni 1597, p. 178c; Mazzoni 2010, p. 275). There is evidence that the *Almagest* was considered a difficult text also in antiquity: see Jones 1999, p. 147.

[17] As an introduction to Theon of Alexandria and his commentary on the *Almagest*, I recommend reading Bernard 2010 and Bernard 2014. See also Juste [83].

commentary on the fifth book and Nicolaus Cabasilas' on the third.[18] According to Barozzi, the *Almagest*

> requires very rich commentaries, such as those composed by Theon of Alexandria, Pappus, and Nicolaus Cabasilas. These commentaries, although erroneous and quite imperfect, can be read in their Greek edition, and – to my knowledge – they have not been translated into Latin by anyone yet.[19]

Barozzi emphasized a significant aspect: during his time, there was no commentary on the *Almagest* that was both accessible to a broader readership—specifically, those who read Latin—and able to provide a comprehensive understanding of its content, including the more complex parts. Barozzi attributed this lack precisely to the challenges intrinsic to Ptolemy's "divine work." In other words, the *Almagest* was hard to read, and therefore required the aid of a commentary, which no one had yet written in Latin due to the difficult reading of the *Almagest*—a sort of catch-22 situation.

> Certainly – continued Barozzi – due to these reasons, no one since Ptolemy has written a complete volume on cosmography, nor has any modern scholar illustrated the divine work of the *Almagest* with commentaries. Instead, all have either written brief annotations on it – like Luca Gaurico, Erasmus Schreckenfuchs, and Erasmus Reinhold –, or have published compendious and elementary instructions on the subject matter, similar to those we have already mentioned.[20]

---

[18] See Halma's introduction in Theon 1821, pp. vi–vii. On the partial dependency of Theon's commentary on that of Pappus, see Bernard 2014, p. 98.

[19] "[…] locupletissimis indiget commentarijs, cuiusmodi sunt ij, qui a Theone Alexandrino, et Pappo, et Nicolao Cabasilla conscripti sunt: qui etiam mendosi, multumque imperfecti Graece impressi leguntur, a neminique nondum (quod ego sciam) Latinitate donati" (Barozzi 1585, f. b5v). See also Clavius 1992, vol. II.1, letter 35, *Barozzi to Clavius* (27 February 1587), p. 74: "[…] dove Theone suo [viz. of Ptolemy] commentatore dice queste parole, le quali metterò Greche per non esser stato esso commentator anchor tradotto […]."

[20] "Quibus profecto de causis, nemo post Ptolemaeum perfectum Cosmographicum volumen scripsit, neque recentiorum quispiam Almagestum ipsum divinum opus commentarijs illustravit; sed omnes aut breves in illud annotationes, ut Lucas Gauricus, Erasmus Osvaldus [viz. Schreckenfuchs], et Erasmus Rheinholt scripserunt: aut Compendiarias, Elementaresque Institutiones eorum, quae in ipso continentur, ediderunt, cuiusmodi sunt, quas iam recensuimus" (Barozzi 1585, f. b5v). In the Italian translation of this passage, printed in 1607, the name of Schreckenfuchs is omitted (see Barozzi 1607, f. 5v).

In the closing lines of this passage, Barozzi refers to the epitome of Proclus and that of Peurbach, which was further developed by Regiomontanus and published posthumously in Venice in 1496. This latter text was certainly crucial for anyone wishing to understand the *Almagest*.[21] Yet, according to Barozzi, the existing annotations and epitomes did not sufficiently clarify the *Almagest*. He even felt that Erasmus Reinhold's annotations on the first book, despite being also based on Theon's commentary, fell short.[22]

In sum, while Barozzi's preface does justify the publication of his *Cosmographia*, being a work capable of filling a gap due to the lack of easy and precise introductions to the *Almagest*, it also underscores the need for a publishing initiative. Such a project would aim at least at making Theon of Alexandria's commentary on the *Almagest* available in Latin, thereby rendering this crucial text more accessible to scholars.

This need had also been felt earlier by Clavius. In the first edition of his commentary on the *Sphere* (1570), he included a lengthy digression on isoperimetric figures, explicitly taken from Theon's commentary on one of the early chapters of the first book of the *Almagest*:

> Moreover, the sphere is greater than all other solid figures with the same perimeter. Although all of these [things] have been geometrically confirmed by Theon, too, in his commentary on Ptolemy's *Almagest*, *since his demonstrations are not readily available to everyone (as only the Greek version [codex] exists), and are indeed very obscure and succinctly demonstrated by him*, I will therefore strive to shed some light on these demonstrations, as much as possible, to satisfy even those who take great delight in geometric demonstrations.[23]

---

[21] See Neugebauer& Swerdlow 1984, p. 51, where the authors also show Copernicus' dependency on the *Epitome of the Almagest*. See also Grafton 1988, p. 789.

[22] On Reinhold's references to Theon, see Omodeo& Tupikova 2013, pp. 241–242, and Omodeo& Tupikova 2018, pp. 21–34. As far as I know, it has not been noted that Reinhold intended to release a second edition of his commentary on the first book of the *Almagest*. This new edition was supposed to include further insights drawn from Theon's commentary. See his final message "to the diligent reader" (*Ad lectorem studiosum*): "Deo iuvante in proxima aeditione hanc institutam explicationem nostram rudiorem perpoliemus, et adiungemus reliqua ex Theone, quae ad Ptolemaei sententiam penitus intelligendam aliquid momenti adferunt. Interea bene vale, et hisce feliciter fruere" (Ptolemy 1549, f. 123r).

[23] "Itemque sphaeram maiorem esse omnibus alijs figuris solidis sibi isoperimetris. Quamvis enim haec omnia a Theone quoque in commentarijs, quos in Ptolemaei

Clavius found it necessary to justify the publication of his treatise on isoperimetry, considering that Theon had already written a similar work complete with geometric proofs.[24] He contended that Theon's work was only accessible in Greek. Furthermore, he found Theon's demonstrations to be "obscure"—a descriptor that Barozzi would later apply to Ptolemy's demonstrations[25]—and developed too quickly, making them challenging to follow. To bridge this publishing void that disadvantaged many, Clavius felt impelled to include an isoperimetry treatise in his commentary on the *Sphere*. This treatise would be republished in subsequent editions of the commentary until 1611.[26] In 1606 it was included in the seventh book of Clavius' *Geometria practica*. Its contents were indeed more consistent with the broader theme of this work. Here, Clavius explains that he had previously taken inspiration from Theon, who had incorporated the treatise into his commentary on the *Almagest*. Furthermore, the version

---

Almagestum composuit, Geometrice sint confirmata; tamen *quia non omnibus in promptu habentur eius demonstrationes, (Graecus enim tantum codex reperitur), et obscure admodum, atque succincte ab eo omnia demonstrantur*; ideo conabor, quod eius fieri poterit, aliquam lucem hisce demonstrationibus afferre, ut vel illis satisfecisse videamur, qui plurimum demonstrationibus Geometricis delectantur" (Clavius 1570, pp. 108–109, emphasis added). The treatise on isoperimetric figures extends over twenty pages (cf. ibid., pp. 109–135).

[24] See Ptolemy & Theon 1538, pp. 11–18 of Theon's commentary.

[25] See *supra*, n. 16.

[26] For instance, cf. Clavius 1581 and Clavius 1585, pp. 81–104, Clavius 1593, Clavius 1602, and Clavius 1608, pp. 96–119, Clavius 1607, pp. 98–123. The treatise is no longer included in Clavius 1611. In the 1581 edition, Clavius writes that he omits other propositions made by Pappus, as he wishes they will soon be published in Latin (Clavius 1581, pp. 80–81: "Caeterarum licet in hoc tractatu solum demonstretur, sphaeram esse maiorem corpore quolibet sibi Isoperimetro, in quo sphaera aliqua describi possit, et quod contineatur vel superficiebus planis, vel conicis, ut suo loco apparebit: Pappus tamen idem de omni corpore demonstravit 70 propositionibus, quas hoc loco apponere supervacaneum duximus, cum brevi, ut spero, Pappus ipse in latinam linguam conversus in lucem sit proditurus"). Evidently, between 1570 and 1581, Clavius was informed by Guidobaldo del Monte that Francesco Barozzi was set to edit and publish Commandino's translation of Pappus' *Collections* (see Clavius 1992, vol. II.1, letter 32, p. 83, and vol. II.2, p. 62, n. 6). By the end of the treatise included in the same 1581 edition, he also hopes an extended version of the treatise will soon be published in a separate work (Clavius 1581, p. 104: "Copiosiorem autem tractationem eadem de re, Deo volente, alio in loco edemus"). In the 1607 edition, Clavius refers to "Pappus Alexandrinum in Mathematicis collectiones" (Clavius 1607, p. 98). Pappus' *Collections* had indeed been published in 1588. Also, in that same edition, on p. 123, Clavius directs readers to the treatise included in his *Geometria practica* (see note below).

of the treatise in *Geometria practica* introduces "three or four [new] propositions."[27]

The significance of Theon's commentary on the *Almagest* to Clavius is manifestly evident in his correspondence.[28] In 1605, a nearly 40-year-old Giuseppe Biancani wrote a letter to his master Clavius, asking him for an occupation involving the translation of Theon[29]:

[...] Brother Giannotti has written to me, saying that you exhort me to become an excellent mathematician. I reply that I have barely interrupted my mathematical studies, except for the time I have spent learning Greek for that project on Theon [*dissegno di Teone*] that you know of. I even recommended Brother Giannotti to you for it, since he knew Greek. Now, I can easily understand the Greek mathematical books [...]. But in the end, what will my studies amount to if you do not assist me someday<?> I fear that one day, seeing myself as useless, I might abandon them. If you would commit to the project on Theon [*dissegno di Teone*], I would offer myself to help, both in comparing the texts with the manuscripts from the Vatican and in translating them, and in doing everything I can. In short, I implore you, and remind you that if my studies are wasted, so are part of yours, and of your efforts.[30]

---

[27] "Quamvis autem de Isoperimetris figuris tractationem bene longam, et copiosam in commentarij nostris in sphaeram instituerimus, exemplum in hoc secuti Theonis Alexandrini, qui idem argumentum in commentarijs in Almagestum Ptolemei persecutus est: tamen quia id in alieno fortassis loco factum esse suspicari quis posset; transferemus eam tractationem ex nostris commentarijs in sphaeram in hanc nostram Geometriam practicam, tamquam in magis proprium locum, additis tribus, aut quatuor propositionibus, quae in illas tractactione desiderantur, et tamen maxime ad hanc materiam spectare videntur" (Clavius 1606, p. 291). On this work, see Little 2022.

[28] As James Lattis emphasized thirty years ago (see Lattis 1994, p. 223, n. 29), it is regrettable that the "monumental effort" made by Ugo Baldini and Pier Daniele Napolitani in editing Clavius' correspondence has never found an appropriate editorial home. Luckily, it is now available online. See https://echo.mpiwg-berlin.mpg.de/content/mpi wglib/clavius (last visited on 26 October 2023).

[29] Note that Galileo and Biancani met in Pauda. On Biancani's mathematical training, see Baldini 2003, p. 73. For bibliographical references on Biancani, see ibid., p. 94, n. 96.

[30] "[...] il F<ratel>lo Giannotti mi scrive che V.R. mi essorta a farmi ott<im>o Math<ematic>o. Rispondo che poco hò interrotto i studii di Mat<hematic>a, se non quanto hò studiato Greco per quel dissegno di Teone che ella sà; et per il quale io già le raccomandai il F<rate>llo Giannotti perche esso sapeva greco: io intendo già i libri di Mat<hematic>a Greci con assi facilità [...]. Ma alle fine à che gioveranno qu<est>i miei studii, se V. R., un giorno non mi dà qualche soccorso <?> Corre pericolo che un giorno,

In previous correspondence between Clavius and Biancani, there is no mention of the "project on Theon" (*dissegno di Teone*). Ugo Baldini and Pier Daniele Napolitani, who edited Clavius' letters, suggested that the two might have discussed it in person during Biancani's stay in Rome, namely between 1598 and 1600. They further suggested that this project was part of a larger endeavor focused on Clavius' commentary on the *Almagest*.[31] Indeed, some of Clavius' correspondents were aware of this project concerning Ptolemy's masterpiece.[32] Those who were not, however, expressed surprise that Clavius had not written a commentary on the *Almagest*. An example is Jacobus Bosgravius, who in 1607 wrote to Clavius expressing such a sentiment:

> I am surprised that your Most Reverend does not write [a commentary] on Ptolemy's *Almagest*. I have never seen an author who interpreted that work, except for Regiomontanus. However, he is hard to understand, both because he is overly concise and because [his epitome] is very poorly printed.[33]

Apparently, Bosgravius ignored the existence of Theon's commentary on the *Almagest*. Or maybe he implicitly referred to "modern" scholars only. At any rate, it is worth noting that his critical view on Regiomontanus' (and Peurbach's) epitome aligns with the earlier mentioned opinion of Barozzi. Furthermore, this critique matches with a passage from Biancani's *Cosmographia* published in 1620, which will

---

vedendo restarmi inutile, non gli abbandoni. Se ella volesse mettersi al dissegno di Teone, io me gli offrirei per aiutante parte in inscontrare i testi con gli manuscritti del Vaticano, parte anche per tradurre, et per fare tutto ciò che io potesse. in somma molto mi le raccomando, et le ricordo che perdendosi li miei studii, si perdono parte de' suoi, et delle sue fatiche" (Clavius 1992, vol. V.1, letter 248, p. 151).

[31] See Clavius 1992, vol. V.2, p. 83, n. 5.

[32] In 1598, Gabriel Serrano wrote a letter from Salamanca to Clavius in Rome asking him to see what he had written on Ptolemy's *Almagest*. Rumors had it that it was published ("*Vehementer videre desidero illos quos in Almagestum Ptolemaei scripsisti, quos fama est, a te in lucem emissos*"). This would have greatly helped him, as he was teaching *Almagest* X.7. See Clavius 1992, vol. IV.1, letter 144, p. 47.

[33] "Miror tuam R<everendissim>am non scribere in Almagestum Ptolemei. Ego nunquam vidi auctorem qui illud opus interpretaretur, praeter Regiomontanum, qui tamen intelligi nequit, quia et compendio studet nimium, et pessime impressus" (Clavius 1992, vol. VI.1, letter 270, p. 58).

be cited later.[34] Therefore, it appears that at that time the *Epitome of the Almagest* was not always viewed as an adequate and self-sufficient supplement to the *Almagest*.

Perhaps Clavius was also aware of this when he planned to publish a commentary on the *Almagest*. Thanks mainly to the studies of Ugo Baldini, it is now possible to get a clearer idea of this project.[35] It was a commentary that, at least in its initial stages, followed the structure of the epitome of Peurbach and Regiomontanus, rather than that of the *Almagest* itself.[36] With this commentary, Clavius aimed to make clear and accessible to students Ptolemy's argumentative process, which was still "obscure" in the *Epitome of the Almagest*, and he intended to do so through a step-by-step explanation—much as he had done with Theon's treatise on isoperimetry and in line with the expository methods followed in his other works. This endeavor extended beyond mere exposition; Clavius sought to incorporate new observational data, hoping to provide a modern perspective on planetary astronomy, which at that time was labeled as "*theoricae planetarum*." While keeping the formal structure of a commentary, Clavius thus aimed to present content that expanded beyond the original lesson of the commented text. His prospective work

---

[34] For Barozzi, cf. the passage *supra*, n. 20, which continues as follows: "Quae si tales forent quod ad omnia in Almagesto contenta nos facile, breviter, ac ordinatim instituerunt, maximam studiosis utilitatem attulissent. Verum omnes iam dictae Compendiariae Institutiones, quas hactenus ego viderim [...] maxima imperfectione patiuntur: cum nulla eorum sit, quae eo, quo decet ordine, facilitate ac brevitate ad omnia, quae Ptolemaeus in Almagesto pertractavit, ad universamque Astrologiam Tirunculos instituat" (Barozzi 1585, ff. b5r–b6v). For Biancani, cf. *infra*, n. 48. That the *Epitome of the Almagest* was "very poorly printed" is a fact already noticed in Neugebauer& Swerdlow 1984, p. 51: "The *Epitome* was printed for the first time in Venice in 1496 in a rather poor edition, and was later reprinted in Basel in 1543 and Nuremberg in 1550. Although it was not served well by its printers, the printing and wide distribution of the *Epitome* was of the greatest importance for all serious astronomy in the sixteenth century, and particularly for Copernicus." See also Pantin 2012, p. 24, n. 30: "In this edition [i.e., the 1496], the diagrams, printed in the margins, are small and often lacking in accuracy." Note that Barozzi, Bosgravius, and Biancani all seem to ignore the existence of George of Trebizond's commentary on the *Almagest*.

[35] See Baldini 1992, pp. 127–145. See also Lattis 1994, pp. 173–179.

[36] See Baldini 1992, p. 134, and Lattis 1994, p. 175.

held the potential to redefine academic standards, challenging esteemed works such as the *Novae theoricae planetarum* by Peurbach.[37]

Although some manuscript parts of this project still exist, it was never completed.[38] It is likely that its core ideas date back to before 1570. This is supported by references in the first edition of Clavius' commentary on the *Sphere*, published in 1570. Other references can be read in subsequent editions, at least until the fifth in 1606.[39] Conceived before the publication of Brahe's works, the project subsequently evolved with the understanding that it would be necessary to also consider the observations of the Danish astronomer.[40] Thus, for more than thirty years, Clavius hoped to write and publish his commentary on the *Almagest*.

This is interesting because it is well-known that Galileo went to Rome in 1587 and there visited Clavius at the Collegio Romano.[41] While it is tempting to imagine their discussions, without evidence, we can only speculate. However, it is already a good thing to be able to place Galileo's visit in a context where the *Almagest* was certainly seen as a text that required a commentary.

---

[37] For a general and recent introduction to Peurbach and his *Theoricae novae*, see Malpangotto 2021, pp. 19–108. See also Aiton 1987, pp. 1–10.

[38] See Baldini's transcription of Clavius' *Theorica Solis* in Baldini 1992, pp. 469–564. According to Baldini, this text was devoted to advanced mathematical education. Yet, it is possible that Clavius would have later simplified it in a manner akin to his other published works. For background on the mathematical Academy of the Collegio Romano between 1533 and 1612, see Baldini 2003.

[39] See Clavius 1992, vol. III.2, p. 69, n. 17; Baldini 1992, pp. 127–129; Lattis 1994, p. 174.

[40] See Clavius 1992, vol. IV.1, *Clavius to Serrano* (21 July 1598), letter 145, p. 55: "Quod scribis de commentariis meis in Almagestum Ptolemaei, scias, me nihil eiusmodi in lucem edidisse. Inceperam quidem ea in re aliquantum laborare, et fere quintum iam librum absolveram, cum ecce rumor affertur, nobilem quendam Danum, Tichonem Brahe, observare denuo motus caelestes, et novas hypotheses moliri." Clavius did also consider Magini's *Theoricae novae* of 1589, which included Copernicus' mathematical advancements. See Baldini 1991, p. 56, and Lattis 1994, pp. 176–179.

[41] See EN, X, *Galileo to Clavius* (8 January 1588), p. 22: "Parmi hor mai tempo di rompere il silenzio sin qui usato con V. S. M. R. da che mi partii di Roma [...]." In Rome, Galileo left a copy of his *Theoremata circa centrum gravitatis solidorum* (see EN, I, pp. 187–208).

The Importance of Theon's
Commentary on the *Almagest*: Biancani

Given this context, it is definitely plausible that Theon's commentary on the *Almagest* would have been highly valuable in relation to Clavius' own commentary (as suggested by Baldini and Napolitani). However, it is also possible that the "Theon project" mentioned by Biancani was an entirely distinct venture, disjointed but complementary to Clavius' commentary on the *Almagest*. In other words, Theon's commentary was considered valuable in its own right.[42]

This, of course, was not news. In the mid-fifteenth century, Cardinal Bessarion provided George of Trebizond with the Greek version of the *Almagest* and a copy of Theon's commentary. Bessarion believed that Theon's insights would simplify the study of the *Almagest*. However, George of Trebizond, who was commissioned by Pope Nicholas V to translate Ptolemy's masterpiece into Latin, argued that Theon's commentary would make understanding the *Almagest* more challenging, contrary to Bessarion's viewpoint. So he wrote his own commentary on the *Almagest*.[43]

This disagreement led to Regiomontanus' *Defensio Theonis*, an unpublished critique largely studied by Michael H. Shank in recent times.[44] This work criticized George of Trebizond's commentary on the *Almagest*, even accusing him of plagiarizing Theon. Furthermore, Regiomontanus'

---

[42] Baldini's most recent statements, in my view, suggest something of the sort. He holds that Clavius' interest, as well as that of some of his students, in Theon's commentary can be attributed to the "influence exerted on Clavius by Italian reconstructions of classical Greek mathematics," as seen in the works of Commandino and Maurolico. See Baldini 2003, p. 61.

[43] See Monfasani 1976, pp. 71, 73–75, 104–108; Monfasani 1984, pp. 671–687, esp. 676–677 for George's critiques against Theon; Shank 2017, pp. 53–65, Shank 2020, pp. 306–308. It is worth noting that, according to Monfasani, "[i]t was not the translation of the *Almagest*, but the commentary which caused Trebizond's fall from papal favor" (Monfasani 1976, p. 113). As far as I know, George of Trebizond's commentary remains understudied.

[44] See Shank 2002, Shank 2007, Shank 2020. Shank and Richard L. Kremer are the editors of a very useful and important digital edition of Regiomontanus' *Defensio*. See https://regio.dartmouth.edu/about/about-project.html (last visited on 26 October 2023). On the *Defensio Theonis*, see also Ekler 2019.

renowned editorial plans included the release of Theon's commentary, possibly with some parts translated into Latin.[45]

However, the 1538 *editio princeps* of the *Almagest* in Greek, inclusive of Theon's commentary, did not revolutionize how astronomy was approached during that era. It seemed to inadequately support an average student's journey in grasping Ptolemy's teachings. This perceived shortcoming, identified by figures like Clavius and Barozzi, was mainly attributed to the complexity of the Greek language. As a result, many believed a Latin translation of Theon's commentary was needed, thinking it could revive Ptolemaic astronomy.[46] Remarkably, this belief persisted at least until the second decade of the seventeenth century.

In this regard, Biancani's recommendations for those pursuing studies in astronomy are worth noting. Although his curriculum, or *cursus studiorum*, for astronomy was published in 1620 and may seem a bit late for our current examination, I believe it provides valuable context. This text is extracted from the second section of Biancani's *Apparatus ad mathematicas addisciendas, et promovendas* (*Apparatus for Learning Mathematics and Advancing its Study*), and is a part of his *Cosmographia*. For clarity, I have segmented the text into six sections, labeled as [a], [b], [c], and so on:

---

[45] See Malpangotto 2008, pp. 46, 103, 191–192, and 151, ll. 20–21. On p. 192, Malpangotto wrongly claims that the 1538 edition of Theon's commentary presented both the Greek text and a Latin translation by Camerarius (the mistake is repeated on p. 215). In note, she refers to Zinner 1990, p. 222, but here there is no mention of a 1538 Latin translation. However, on the same page, she interestingly notes that "[i]l catalogo più antico della biblioteca di Norimberga, redatto postumo nel 1512, include anche: 'Theonis traduction [*sic*] Conclusiones', che sembra segnalare una traduzione dell'opera di Teone, probabilmente eseguita da Regiomontano, della quale però non si hanno notizie."

[46] The mismatch between observational data and the results from astronomical tables served as a first powerful catalyst for the revitalization of Ptolemy's astronomy in Europe, especially evident in the second half of fifteenth century. See Malpangotto 2008, pp. 11–12, 58. In the fifteenth century, one solution (advocated by Bessarion and then Regiomontanus) appeared to be a return to the original Greek sources. However, by the second half of the sixteenth century, emphasis was placed on producing Latin translations to make the now-accessible Greek texts, such as Theon's commentary, more widely available. Scholars like Clavius might have hoped that this approach would prove advantageous in discovering "new geocentric schemes congruent with the traditional measurements." However, the landscape transformed entirely after the publication of Brahe's data (see Baldini 2003, p. 57).

[a] First, attention should be given to the treatises on the *Sphere*; however, unless self-love deceives me, I advise you to first read this *Sphere* of mine. Then, because of its antiquity, one should read the *Sphere* of Proclus Diadochus; Cleomedes' *Meteorology* translated by Valla; Euclid's *Phaenomena* translated by Auria; Campanus's *Sphere*; Maurolico's *Cosmography*; al-Farghani's *Rudimenta astronomica* translated by Jacobus Christmannus; [Alessandro] Piccolomini's Italian *Sphere*; Clavius's *Sphere*; Autolycus, *On the Moving Sphere*; of the same author, *On Risings and Settings*, translated by Josephus Auria; Theodosius of Tripoli, *On Days and Nights* and *On Habitations*, translated by Auria; Julius Hyginus on the sphere and celestial signs.[47]

[b] Next should come the epitome of Johannes Regiomontanus on Ptolemy's *Great Construction*; it is indeed obscure but can be understood with the help of Geometry and the doctrine of plain and spherical triangles (about which, see [the works cited] above), even though the faulty printing is a significant hindrance. Then comes Ptolemy's *Great Construction*, in Arabic the *Almagest*, a wondrous work containing all of Hipparchus and Ptolemy's astronomy. There are two translations of it: one from Arabic [by Gerard of Cremona] and the other from Greek by George of Trebizond. There is a Greek commentary on the latter by Theon, which can be useful for those who know Greek, as it has been published. Proclus made a summary of [the *Almagest*] under the name *Hypotyposis of Astronomical Positions*. The Arabic Geber's astronomical work is like the *Almagest*; in fact, it explains the *Almagest* and frequently corrects Ptolemy. Al-Battani, or Muhammad of Raqqa's *De scientia stellarum* should be read in parallel. After him comes Nicolaus Copernicus, who, besides his absurd hypothesis about the motion of the earth, is an excellent astronomer. However, now he should be read with the permission of the Church.[48]

---

[47] "Primum studium impendatur tractatibus de Sphaera; ego autem ni philautia me fallita, Sphaeram hanc meam tibi primo legendam consulo: deinde ob antiquitatem legatur sphaera Procli Diadochi, Cleomedis meteora Valla interprete; Euclidis Phaenomena Auria interprete. Campani sphaera, Maurolyci Cosmographia, Alfragani Elementa Astronomica, ex traductione Iacobi Christmanni, Sphaera Piccolomini Italica, Clavij sphaera, Autolycus de sphaera quae movetur; item de vario ortu et occasu astrorum, Iosepho Auria interprete, Theodosius Tripolita de diebus, ac noctibus, item de Habitationibus Auria interprete, Iulius Higinius de sphaera, ac signis caelestibus" (Biancani 1620, p. 396).

[48] "Deinceps succedat Epitome Ioan. de Monteregio in Magnam Ptolemaei constructionem; est quidem obscura sed tamen auxilio Geometriae et doctrinae de triangulis planis, et sphaericis, de quibus supra, adhibitio studio intelligi potest, quamvis multum obest impressio mendosa. Deinde Ptolemaei Magna constructio Arabice Almagestum, opus mirum, et in quo tota Astronomia Hipparchi, et Ptolemaei continetur. Duae ipsius versiones circumferuntur, una ex Arabica lingua, alter ex Graeca Georgij Trapezuntij,

[c] Finally, the works of Tycho Brahe, namely, *Progymnasmata, On Comets*, [*Astronomiae instauratae*] *Mechanica, Epistolae astronomicae*, should conclude this study.[49]

[d] Afterwards, these can be approached without any order: Johannes Kepler on the new star in Ophiuchus, and on another new star in Cygnus; Galileo's *Sidereus Nuncius*, and his [letters] on sunspots in Italian; Apelles behind the picture [i.e. Scheiner] on the same topic; Guidobaldo's *Astronomical problems*; Aratus of Soli's *Phaenomena* sung in verse—there are commentaries on it by Hipparchus in Greek, and published in Latin; translations in Latin verse of the same Aratus by Germanicus Caesar, Cicero, and Rufus Avienus can be read for education. Also, Hipparchus on the constellations published with the aforementioned commentary. Archimedes' *The Sand Reckoner* seems to be an astronomical book to be placed here. Alhacen's and Witelo's *De Crepusculis*; various works by Pedro Nunes.[50]

[e] However, authors of *theoricae planetarum* seem to me full of pointless labor and imagination. In lieu of the *theoricae*, the works already mentioned are more than sufficient.[51]

[f] Other works not yet published will be listed [*infra*] in their own catalog.[52]

In the first four sections, from [a] to [d], Biancani lists the works useful for acquiring a scientific proficiency in theoretical astronomy in his time.

habentur in ea commentaria Theonis Graeca, quae graece scienti auxilio esse poterunt, sunt enim edita. Huius compendium fecit Proclus sub nomine Hypotyposis Astronomicarum positionum. Geber arabs opus astronomicum est instar Almagasti [*sic*], imo Almagastum [*sic*] exponit, et passim Ptolemaeum redarguit. Albategnius, sive Mahometes Aratensis de scientia stellarum nunc pariter legendus. His succedat Nicolaus Copernicus, qui praeter absurdam hypothesim de motu terrae eximius est Astronomus, sed nunc cum Ecclesiae facultate legendus" (ibid., pp. 396–397).

[49] "Tandem opera Tichonis Brahe, id est, Progymnasmata, de Cometis, Mechanica, Epistolae, claudant hoc studium" (ibid., p. 397).

[50] "Postea sine ordine adiri possunt hi, Ioan. Keplerus de Stella nova in Serpentario, Idem de alia nova in Cygno; Galilaei nuncius sydereus, idem de Maculis solaribus italice; Apelles post tabulam latens de ijsdem maculis; Guidiubaldi problemata astronomica; Arati solensis phaenomena versibus decantata: extant in ea commentaria Hypparchi Graece, et Latine edita: Germanici Caesaris, Ciceronis, et Rufi Avieni translationes versibus latinis eiusdem Arati legi possunt ad eruditionem. Item Hypparchus de Asterismis editus cum praedicto commentario. Archim. de Arenae numero, videtur liber Astronomicus hic collocandus. Alhazen et Vitello [*sic*] de Crepusculis; Petri Nonij varia" (ibid.).

[51] "Authores vero Theoricarum Planetarum videntur mihi inani laboris et imaginationis pleni, pro quibus sufficiant abunde recensiti" (ibid.).

[52] "Alij nondum editi in Catologo proprio scribentur" (ibid.).

It was essential to start with a solid knowledge of spherical astronomy. As pointed out in section [a], this basic knowledge was best acquired from the latest comprehensive work, specifically Biancani's own *Cosmographia*. Following this, students were expected to delve into classic texts, ranging from ancient to contemporary, all predominantly in Latin, with a notable exception being Piccolomini's *Sphere* written in vernacular.

Once this foundational knowledge was established, students could then engage with Ptolemy's *Almagest*, the centerpiece of section [b]. This section also introduced them to Copernicus' *De revolutionibus*. But before diving deep into these, Biancani recommended the epitome by Regiomontanus, even with its known imperfections.[53] Furthermore, before fully embracing Copernicus, students should also explore Proclus' *Hypotyposis* and two pivotal Arabic works: Geber's *Liber super Almagesti* and al-Battani's *De scientia stellarum*.[54] Only with this background could students fully appreciate Copernicus. However, reading Copernicus' *De revolutionibus* was conditional, requiring ecclesiastical approval due to the 1616 anti-Copernican decree.[55]

Section [c] wrapped up the curriculum with a nod to Brahe's contributions, which, at that time, were becoming pivotal for Jesuit astronomers.[56]

---

[53] See *supra*, nn. 21, 34.

[54] See Geber 1534 and Juste's entry on this work in Juste [70]. As for al-Battani's *De scientia stellarum*, it is worth noting that, thanks to Johannes Schoener, at that time circulated also a printed version with Regiomontanus' additions: see al-Farghani& al-Battani 1537.

[55] See Pagano 1984, pp. 102–103; Finocchiaro 1989, pp. 148–150; Pagano 2009, pp. 46–47.

[56] As Luís Miguel Carolino has recently recalled, "[a]fter a distressing process of censorship, Giuseppe Biancani's *Sphaera Mundi* was finally published in 1620. Although Biancani's book was to a large extent just a traditional treatise on cosmography, it was nevertheless the first printed work by a Jesuit author to endorse the Tychonic planetary system. For such reason, it has become regarded as a turning point in the science politics of the Jesuits, the point in time when the Jesuit authorities officially accepted the Tychonic geo-heliocentrism" (Carolino 2023, pp. 2–3). On this, see the important seminal work Lerner 1995, esp. p. 178: "Entre 1620, date de la *Sphaera* de Biancani, et 1651, quand paraît l'*Almagestum novum* de Riccioli, l'hypothèse de Brahe a acquis droit de cité: Riccioli écrit que la majorité des jésuites ont choisi ce système dans leurs traités *De caelo*." Concerning the censorship of Biancani's *Cosmographia*, see documents and notes in Baldini 1992, pp. 232–238, 245–250. See also Blackwell 1991, pp. 148–153.

Section [d], although touching upon the contributions of Kepler and Galileo—among others—, was more supplemental, serving as an addition rather than a foundation.

Interestingly, in section [e], Biancani opted to leave out Peurbach's *Theoricae novae planetarum* and its accompanying commentaries, a departure from the academic convention of the time. Conventionally, after grasping spherical astronomy, students progressed to the *Theoricae* and then to the *Almagest*.[57] However, Biancani believed that with just Regiomontanus' epitome, students could approach the *Almagest*, bypassing the *Theoricae*. For those well-versed in Greek, Theon's commentary on the *Almagest* offered a richer understanding.

This approach matches with what was said earlier regarding Clavius' project. A proper commentary on the *Almagest*, or rather on Regiomontanus' (and Peurbach's) epitome, would have rendered the preliminary study of Peurbach's *Theoricae novae* unnecessary. However, in the absence of such a commentary, Theon's commentary was seen by Biancani as a pivotal educational tool, effectively overshadowing Peurbach.

Interestingly, in section [f], Biancani hinted at works awaiting publication. A few pages later, in the fourth section of the *Apparatus*, he listed several important mathematical works that were either unpublished or not yet translated into Latin. The concluding entry is Theon's commentary on the *Almagest*:

> Lastly, there remains the Greek commentary of Theon of Alexandria on Ptolemy's *Great Construction*. It has indeed been published, but not yet translated into Latin, even though *many have attempted this translation, and* [despite the fact that] *many also write that they especially desire it.*[58]

---

[57] See Lattis 1994, p. 41. See also what Mazzoni says about it: "Tractatus ergo spherae et planetarum theorica paediam continent praeparantem ad magnam Ptolomaei compositionem" (Mazzoni 1597, p. 178d–e; Mazzoni 2010, pp. 275–276). Castelli and Galileo recommended a similar curriculum to Ludovico Delle Colombe so he could read, understand, and consequently agree with Copernicus (see Malara 2023, p. 471, and *supra*, Chapter 1, n. 1).

[58] "Ultimo restant Theonis Alexandrini Graeca commentaria in Ptolemaei Magnam constructionem, edita quidem, sed nondum in latinum translata quamvis *multi eam translationem sint aggressi, et multi etiam eam se maxime desiderare scribant*" (Biancani 1620, p. 405, emphasis added).

As already seen, among those who wished for a Latin version of Theon's commentary on the *Almagest* were Clavius and Barozzi. The first to partially exploit this publishing void was Giovanni Battista Della Porta.[59] In 1605, he released a Latin version of the first book of the *Almagest*, accompanied by Theon's pertinent commentary. Although late sources mention a possible 1588 edition, no copies or contemporary references to this alleged first edition have come down to us. With our present understanding, the existence of a publication preceding the 1605 edition seems highly improbable.[60]

And yet the "desire" for a comprehensive Latin translation of Theon's commentary did not dissipate after 1605. As Biancani noted in 1620, Della Porta was just one among "many" who sought to translate that commentary. In 1610, for example, Clavius was notified by Giovanni Giacomo Staserio about the existence of a Latin manuscript of Theon's commentary in Naples, and that one of his students was considering publishing it.[61] In fact, several manuscript translations were circulating in Europe during the latter half of the sixteenth century.[62] This corroborates, once again, that there was a strong demand for a Latin commentary

---

[59] For a short overview on Della Porta's life and work, see Verardi 2015. Note that Della Porta became a Lyncean in 1610, just a year before Galileo. Probably the two met in Padua in 1593.

[60] See Raffaella De Vivo's introduction in Della Porta 2000, pp. XIV–XV, where she also suggests other hypotheses on the 1588 alleged edition.

[61] See Clavius 1992, vol. VI.1, letter 317, p. 156. Staserio incorrectly credited the translation to Pedro Nunes. Both Baldini and Napolitani observed this mistake and, thanks to Romano Gatto, deduced that Staserio was referencing Jerónimo Muñoz (see ibid., VI.2, p. 93, n. 5). Unfortunately, Staserio's attempt to publish this translation was unsuccessful: "Scrissi della tradottione de commen<ta>rii di Theone s<opr>a l'Almagesto, come stavo per haverlo, adesso li dico, come non si sono avuti. un gentilhuomo Genovese voleva pagarli 60 ducati con darli insieme parola di farli stampare, ma chi l'havea per sua dapoca<gin>e non ha saputo conoscere tanta buona sorte. credo che saria stato un buon libro per stampa, ma di gran fatica" (Clavius 1992, vol. VI.1, *Staserio to Clavius* (31 December 1610), letter 320, p. 163). Muñoz's manuscript has been partly transcribed by Miguel Á. Granada (see Granada 2023). The manuscript is now fully available online via the *Ptolemaeus Arabus et Latinus* website (see Juste [164]). On Muñoz, see Navarro Brotons 2019.

[62] Thanks to the *Ptolemaeus Arabus et Latinus* website, today it is really easy to read the digital copy of some of these handwritten Latin translations scattered throughout Europe. See: https://ptolemaeus.badw.de/filter?text=theon+of+alexandria (last visited on 26 October 2023).

on the *Almagest* during that period. More specifically, there was a strong demand for a Latin translation of Theon's commentary.

A Latin Translation of Theon's Commentary on the *Almagest*: Teofilo

Focusing on the early Galileo and the environments closely related to him, at least two manuscript translations deserve mention. One, dating back to around 1571, was penned by Giovanni Battista Teofilo, a physician and mathematician from Urbino, of whom we know little.[63] The other, an anonymous work, is held in Florence (Biblioteca Medicea Laurenziana, Acquisti e Doni 694) and was copied in the final years of the sixteenth century.[64] It is unclear how widely these translations circulated.

Here, I will focus on Teofilo's version as there is more literature available on it, highlighting the relevant evidence. In 1971, Paul Lawrence Rose released an article where he transcribed two original letters from Federico Commandino, discovered at the Bibliothèque nationale de Paris.[65] Both letters were written to Teofilo. In the second one, dated July 30, 1574, Commandino explicitly wanted

> to know if Your Lordship [Teofilo] has completed the translation of Theon, and the other commentators [i.e., Pappus and Cabasilas] of the *Almagest*, or where you stand with it.[66]

And, he appended in the postscript:

> Trapezuntius authored a commentary on the *Almagest* after translating it, and one could assume that he translated those other commentators [i.e., Theon, Pappus, and Cabasilas] as well, incorporating them into his commentary. It might be worth looking at it, considering it is housed in

---

[63] See Juste [85]. Along with Guidobaldo del Monte, Bernardino Baldi, Valerio Spaccioli, Alessandro Giorgi, Felice Paciotti, Francesco Corona, and likely others, Teofilo was a pupil of Commandino: see Gamba& Montebelli 1988, p. 24, n. 21.

[64] See Juste [165].

[65] See Rose 1971.

[66] "[…] intendere se Vostra Signoria finì di tradurre Theone, et gli altri commentatori dell'Almagesto, o in che termine se sia" (ibid., p. 307).

the library of Saint Peter, and with some favor, Your Lordship might be able to have it brought out at your leisure.[67]

It is also from letters like this that one can grasp the significant role that commentaries on the *Almagest* were perceived to have in the sixteenth century. Even George of Trebizond's commentary, despite being heavily criticized by Regiomontanus and remaining unpublished, held potential value in Commandino's view.[68] This was particularly due to the hope that it might contain excerpts from Theon's commentary. Perhaps Commandino was aware that George of Trebizond had been accused, precisely by Regiomontanus, of plagiarizing Theon's commentary? Regardless, he urged Teofilo to complete his translation.

Bernardino Baldi also acknowledged Teofilo's work in his *Life of Ptolemy*:

The work of Theon on this author [that is, Ptolemy] was translated into Latin years ago by Giovanni Battista Teofilo, a physician and my fellow countryman. A man of divine intellect and a devoted mathematician, Teofilo could not put the final touches on his work, as he was hindered by the course of human tumult and cut short by death. The work is now in the hands of his nephews who, as I understand it, are striving to benefit the world and illuminate their uncle's memory by publishing it.[69]

This translation, however, was never published. Despite this, Teofilo's nephews endeavored to preserve their uncle's memory and eventually passed his work to Galileo's last student, Vincenzo Viviani, many years later. Viviani subsequently handed the manuscript to Jean Mabillon. In

---

[67] "Il Trapezuntio fece un commento sopra l'Almagesto havendolo tradotto, et si può credere che habbia tradotto ancora quei commentatori, et gli habbia posti nel suo commento, forse non saria se non bene di vederlo, perché essendo nella libraria di San Pietro Vostra Signoria con qualche favore lo poscia havere fuori a suo piacere" (ibid.).

[68] See *supra*, nn. 43 and 44.

[69] "L'opera di Teone sopra questo autore fu tradotta in latino gl'anni passati da G. Battista Teofilo Medico, e mio compatriota, huomo d'ingegno divino, e studiosissimo delle Matematiche ma non pote dargli l'ultima mano impedito dal corso delle turbolenze humane e prevenuto dalla morte; l'opera è in mano di nipoti daquali per quanto intendo si procura di far utile al Mondo e d'illustrar la memoria del zio col pubblicarla" (BB, f. 340r). A reference to this passage can also be found in the *Life of Teone* (autograph): see BB, f. 298r.

the *Iter Italicum litterarium annis MDCLXXXV et MDCLXXXVI*, Mabillon indeed wrote that on April 28, 1686,

> [Viviani] handed into our hands the translation of Theon's [commentary] on Ptolemy, made in the previous century by Giovanni Battista Teofilo, a nobleman from Urbino, to be transported to the royal library. We delivered this to the most illustrious Thevenot, curator of the royal library.[70]

In an act of handover, Viviani presented Teofilo's translation to Mabillon, who then passed it to Melchisédec Thévenot, the curator of the French royal library since 1684. This sequence of events is confirmed (although with a slight difference in date) by a note Viviani penned himself on the last page of the manuscript (Fig. 2.1):

> On the first of May <u>1686</u>
> In Florence
> I, Vincenzo Viviani, mathematician of the Serenissimo Grand Duke of Tuscany, freely present this manuscript to the library of His Most Christian Majesty [i.e., Louis XIV]. The manuscript contains a translation from Greek of Theon of Alexandria's commentary on Ptolemy's *Almagest*, completed before <u>1571</u> by Giovanni Battista Teofilo, a nobleman from Urbino, and a highly learned physicist and mathematician, who was a contemporary of Federico Commandino. I obtained this manuscript 14 years ago from the heirs of the aforementioned Teofilo
> I, the aforementioned, by my own hand.[71]

---

[70] "Idem [viz. Illustrissimus Vivianus] nobis Theonis in Ptolemaeum versionem, saeculo superiori a Johanne Baptista Theophilo, nobili Urbinate, factam, ut ad regiam bibliothecam transportaretur, tradidit in manus: quam clarissimo Thevenoto regiae bibliothecae curatori consignavimus" (Mabillon 1687, p. 193). This is also recalled by Giacomo Leopardi in his early *History of Astronomy*.

[71] "A di primo Maggio <u>1686</u> / in Firenze / Io Vincenzio Viviani matematico del Se<renissi>mo Granduca di Toscana fò libero dono alla biblioteca di S<ua>M<aes>tà Crist<ianissi>ma del presente manoscritto contenente la Versione dal Greco de' Comenti di Teone Alessandrino sopra l'Almagesto di Tolomeo fatta avanti al <u>1571</u> da Gio<vanni> Batt<ist>a Teofili nobile Urbinate e dottissimo fisico e matematico; e coetaneo di Federigo Commandini; il qual manoscritto ottenni son già 14 anni dagli eredi di d<etto> Teofili / Io sud<detto> manu p<ro>p<ri>a" (GBT, f. 383r, underlined in the original manuscript, accessible online on the *Ptolemaeus Arabus et Latinus* website via the following link: https://ptolemaeus.badw.de/ms/113/175/383r).

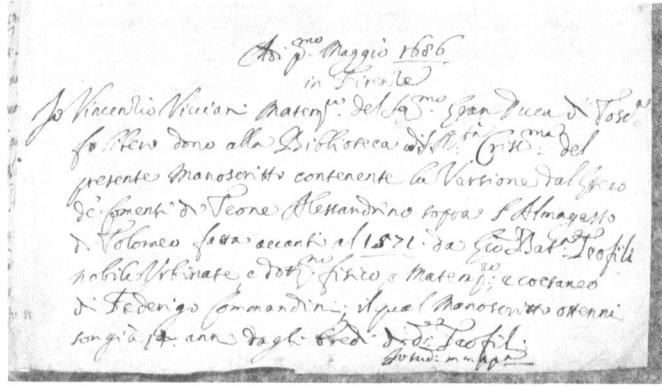

Fig. 2.1   Paris, Bibliothèque Nationale de France, Latin 7263, f. 391. (Courtesy of the BnF)

This handwritten note indicates that Viviani obtained the manuscript in 1672, a full thirty years after Galileo's passing. Thus, there is no evidence to suggest that Galileo had the chance to read any parts of Teofilo's translation of Theon's commentary on the *Almagest*. Yet, equally, there is no substantial evidence to rule out this possibility. The matter remains open for investigation, pending a more thorough study of the circulation and impact of Teofilo's translation, and others like the anonymous Florentine version mentioned earlier.

Knowledge of Theon's Commentary on the *Almagest* at Pisa: Buonamici and Mazzoni

It is worthwhile to note that in the academic milieu of Pisa, there were individuals capable of reading Greek, and at least two of them engaged with Theon's commentary.

Mario Otto Helbing noted that from certain excerpts of Francesco Buonamici's *De motu* (1591), it is apparent that Buonamici had consulted the *editio princeps* of 1538, specifically Theon's commentary. For instance, Theon's commentary is used in the fifth book of *De motu* to add an argument against Copernicus' heliocentric thesis.[72] Buonamici, who was a professor of natural philosophy in Pisa, had Galileo as one of his students.[73]

Moreover, Jacopo Mazzoni refers to Theon's commentary in his renowned 1597 work *De comparatione*.[74] While discussing the hexagonal form of beehives, he relies on what Theon demonstrated in his commentary on the first book of the *Almagest*:

> Therefore, driven by nature's instinct, they [the bees] strive to create their work solid, and not wavering, and as capacious as possible. And because among other shapes that leave no empty space, the hexagon has more corners, and therefore (as Theon proved in his commentaries on Ptolemy's *Almagest*) it is more capacious, hence, for a larger honey production, the bees selected the hexagonal shape among those three [i.e., the hexagon, the square, and the triangle].[75]

Mazzoni taught philosophy in Pisa from 1588 to 1597 and was both a colleague and friend of Galileo. Galileo mentioned in a 1590 letter to his

---

[72] See Helbing 1989, pp. 195–197, 266 n. 13. See also pp. 47 and 225 where Theon is included among the sources of Buonamici. On p. 196, Helbing says that Buonamici referred to Theon's commentary on Book VII of the *Almagest*. This is evidently a typo, as on p. 197 n. 5 Helbing correctly refers to *Alm*. I.7 in Rome's edition. Luigi Guerrini does not see the typo and repeats it in his book on the early Galileo (see Guerrini 2011, p. 183). To the passages already highlighted by Helbing, one can add Buonamici 1591, p. 458e, where Eratosthenes' digit of 10 stadia is taken directly from Theon's commentary on *Alm*. I.4 ("… ut docet Theon ex Eratosthene summorum montium altitudo non auget diametrum terrae plusquam spatio 10 stadiorum…"). See Ptolemy & Theon 1538, Theon's commentary, p. 23, ll. 8–9, Ptolemy & Theon 1605, p. 41, Della Porta 2000, p. 62, ll. 265–266. For background on this passage from Buonamici's *De motu*, see Helbing 1989, pp. 201–203.

[73] See Camerota 2016.

[74] On this work and Mazzoni's relationship with Galileo, see Purnell 1971 and Purnell 1972.

[75] "Instinctu itaque naturae opus suum, tum solidum, et minime mutans, tum maxime capax, efficere conantur, et quia inter alias figuras, quae nihil vacui reliquunt, exagona est pluribus angulis instructa, et proinde ex illis, quae Theon in commentarijs ad magnam Ptolemaei compositionem probavit, capacior est, ideo ad ampliorem mellificationem inter illas tres figuras exagonam Apes seligerunt" (Mazzoni 1597, p. 195A–B; Mazzoni 2010, p. 300). In fact, this excerpt bears a striking resemblance to the beginning of the fifth book of Pappus' *Mathematical Collections* (see Pappus 1588, ff. 73r–74r). I express sincere gratitude to Eileen Reeves for bringing this to my attention.

father that he was "studying and learning from Mr. Mazzoni."[76] While it is not entirely clear what Galileo was studying and learning, it is plausible that they discussed the motion of heavy bodies among other topics.[77]

At this point, it would be tempting to suggest that they did also discuss Ptolemaic astronomy together. In the past, there have been speculations based upon a letter Galileo sent to Mazzoni on May 30, 1597. This letter is significant as it is one of the earliest instances in which Galileo openly aligns with the Copernican view. In *De comparatione*, Mazzoni criticized the heliocentric model proposed by Copernicus, using what I would call a 'variation' on the horizon argument. The traditional stance posited that if the Earth were not at the center of the universe, our horizon should not bisect the celestial sphere evenly, yet that is precisely the phenomenon we observe.[78] Mazzoni argued that if one were on top of a towering mountain, such as the Caucasus, the horizon would not evenly bisect the celestial sphere. So much so, the notion of the Earth being distant from the center of the world, as Copernicus proposes, seems even more preposterous.[79] In response, Galileo emphasized that there is a much greater observational difference in viewing the celestial sphere from a mountain's peak than from an Earth that, according to Copernicus, is situated away from the center of the universe.[80]

Without delving into extensive detail, it has been suggested in historical accounts that Mazzoni and Galileo also engaged in debates over Ptolemaic and Copernican astronomy. For instance, Luigi Guerrini, drawing on conjectures made by Wallace and Drake, imagined that "one of the topics of contention between the two colleagues [Galileo and Mazzoni] in Pisa might have centered on the motion of the Earth."[81] However, these

---

[76] EN, X, 44–45.

[77] See Camerota& Helbing 2000, pp. 344, 364.

[78] More on the horizon argument *infra*, Chapter 3.

[79] "Si montis Caucasi altitudo tanta est ut tempore solstitii aestivalis […] quicunque in eius apice esset per tres horas, si coniunctim assumatur utrumque crepusculum, et per sex, si divisim, Solem videret, et per consequens de duodecim Zodiaci signis, aut septem cum dimidio alterius, aut novem integra, supra terram aspiceret, proculdubio altitudo centri terrae supra centrum mundi, secundum Copernicum multo maior quam situ altitudo Caucasi, non potest esse adeo insensibilis, ut ex utraque parte aequalem nobis obtrudat mundi portionem" (Mazzoni 1597, p. 133S; Mazzoni 2010, p. 210).

[80] See EN, II, *Galileo to Mazzoni*, pp. 197–202. See also Bellone 2003, pp. 31–35.

[81] Guerrini 2011, p. 194: "Le due cose [i.e., the fact that in the aforementioned letter, Galileo refers to his discussions with Mazzoni in Pisa, and then rejects Mazzoni's

conjectures cannot be checked against documents currently available to us.

To avoid proposing conjectures that, while intriguing, are not entirely checkable, I must narrow my focus to a specific aspect. It seems to me that Galileo's refutation of the 'variation' on the horizon argument had not been explained by Galileo to Mazzoni during their tenure as colleagues in Pisa, otherwise he would have not repeated it in his 1597 letter. This could suggest that Galileo remained persuaded by the Ptolemaic arguments regarding this issue at that time. Alternatively, it might indicate that they did not delve into thorough discussions concerning Ptolemaic and Copernican astronomy.

CONCLUSION

Between the sixteenth and seventeenth centuries, in North-Central Italy, there was a pressing need to provide comprehensive commentaries for the *Almagest*. This was driven partly by educational demands and partly by the desire to revitalize the waning Ptolemaic astronomy. Clavius envisioned a commentary that would both simplify and modernize the study of the *Almagest*. Perhaps it would have culminated in an encyclopedic work like Giovanni Battista Riccioli's *Almagestum novum* (1651), but this must remain uncertain.[82] It was undoubtedly a challenging project to execute. Many scholars shared a more feasible goal: printing a complete Latin translation of Theon's commentary on the *Almagest*. A number of handwritten translations and Della Porta's printed translation of the first book show that this project was indeed feasible, yet it was never realized.[83]

'variation' on the horizon argument], così consecutivamente accostate, fanno in modo legittimo immaginare che uno degli argomenti delle controversie avvenute a Pisa fra i due colleghi potesse essere stato proprio il moto della terra." See also p. 195 and ff. for references to Wallace and Drake.

[82] See Baldini 1992, p. 140, partially translated in Lattis 1994, p. 176. It is worth noting that besides Clavius, there is evidence that Ercole Bottrigari was also working on a commentary on the *Almagest* (see Pellacani 2020, esp. pp. 54–65).

[83] Even now, we still lack a critical edition and a modern language translation of Theon's entire commentary. Nicholas Halma translated into French the first two books of the commentary (see Theon 1821). Adolphe Rome completed the critical edition of Ms. Biblioteca Medicea Laurenziana, *Plut.* 28, 18, which includes the oldest surviving copies of Theon's and Pappus' commentary on *Almagest* I-VI (see Theon 1931–1943).

Bearing this context in mind, it is uncertain whether Galileo, around 1590, had read sections of Theon's commentary—either in Greek, with Mazzoni's assistance, or in Latin, by referring to a handwritten translation.[84] He had likely heard of it, though.

In any case, Galileo's commentary on the *Almagest* would have been a natural fit within the publishing environment of the time. It almost appears as though he had recognized this literary void and sought to address it. He certainly seemed to consider himself well-equipped to undertake such a challenging task.

---

[84] Galileo was not completely unfamiliar with Greek. However, the prevailing belief among scholars, with only a few dissenting (e.g., see De Pace 2005, and De Pace 2020, pp. 355–395), is that Galileo possessed merely a rudimentary understanding of the Greek language. This limited knowledge would not have permitted him to independently interpret a Greek text. For more information on this, one can refer to Romana Berno 2008, pp. 19–20 nn. 14–19. In the present study, in an effort to steer clear of making baseless speculations, I choose to align with the perspective that has gained broad acceptance.

CHAPTER 3

Disentangling the Ptolemaic
from the Aristotelian

Abstract The third chapter examines Galileo's references to Ptolemy in *De motu antiquiora*, focusing on Chapters 5–7 of the *Almagest* I. Galileo acknowledges Ptolemy's importance and accepts his geocentrism, distinguishing it from Aristotle's views in *De caelo*. However, he critiques Ptolemy's failure to explain the Earth's central position in the world, offering an alternative rationale influenced by Archimedes' hydrostatics. This chapter highlights how Galileo integrates Ptolemy's arguments into a new cosmological perspective informed by Archimedes' theories.

Keywords Galileo Galilei · Claudius Ptolemy · *Almagest* · *De motu antiquiora* · Archimedes · Aristotle · Geocentrism

In April 1607, Benedetto Castelli was at the Abbey of the Holy Trinity in Cava de' Tirreni, located in southern Italy. From there, he exchanged letters with Galileo, who was residing in Padua at the time, having moved there at the end of 1592. Their correspondence reveals that they had met

© The Author(s), under exclusive license to Springer Nature Switzerland AG 2024
I. Malara, *Galileo and the* Almagest, *c.1589–1592*, Palgrave Studies in the History of Science and Technology,
https://doi.org/10.1007/978-3-031-70614-1_3

some time earlier in Padua, where Castelli was introduced to mathematics by Galileo himself.[1]

At the beginning of his first letter from Cava de' Tirreni, Castelli updated Galileo on his mathematical studies. After delving into Euclid's *Elements*, he began studying the *Almagest* but was soon baffled by a challenging "corollary" from Chapter XII of the first book. In search of clarity, he reached out to Galileo, considering him a known expert on the topic.[2] Indeed, Galileo had taught the *Almagest* at the University of Padua.[3] But it is likely that his knowledge of Ptolemaic astronomy began earlier during his teaching days in Pisa. Records from Pisa, like the Pisan *rotuli* and the *De motu antiquiora* manuscripts, support this theory.[4]

Before delving into *De motu antiquiora*, three important premises should be addressed. Firstly, Ptolemy and the *Almagest* are also mentioned in other early writings by Galileo, namely the so-called *Juvenilia*.[5] This collection, comprising traditional teachings copied from

---

[1] See EN, X, *Castelli to Galileo* (1 April 1607), pp. 169–171. Regarding the details of Castelli's education before meeting Galileo, one can only speculate. However, it has been shown that Benedictine training in Brescia, Castelli's birth city, and Padua was mainly centered on theology. See Piccinali 2018, pp. 49–121. Massimo Bucciantini highlighted that in the Benedictine Abbey of Santa Giustina in Padua, where Castelli resided, there was a strong Neoplatonic influence that echoed in discussions on the "ruolo e il grado di certezza delle matematiche [che] si intrecciavano a quelle sui rapporti tra matematica e teologia, tra simbologia cristiana e figure e simboli geometrici" (Bucciantini 1992, p. 173).

[2] "Dopoi ho dato l'assalto a Tolomeo, ma son restato intricato al primo corollario del capitolo duodecimo: se V. S. mi vole favorire con darmi qualche lume, infilzarò quest'obbligo con gli altri" (EN, X, *Castelli to Galileo* (1 April 1607), pp. 169–170). Castelli here refers to *Almagest* I.12 (that is, I.13 in Toomer's translation). None of the Latin translations of the time, including the one by Della Porta with Theon's commentary, mentions a corollary in this chapter. Here, Ptolemy deals with spherical trigonometry. He lays out four "preliminary theorems" or "lemmas," and from them, he deduces two other theorems (see Toomer 1984, pp. 68–69). It may be that Castelli called the first of these last theorems derived from the previous ones a "corollary." It must be noted, however, that the term "correlarium primum" is found in one of Luca Gaurico's marginal additions in the ninth chapter of the first book, as included in the Latin edition of 1528: "Correlarium primum: Data alicuius arcus chorda, nota fiet chorda arcus residui de semicirculo" (Ptolemy 1528, f. 5r).

[3] He lectured on "Almagestum Ptolomei" in 1597–1598 (see EN, XIX, p. 120 n. 1). It must be noted that he also lectured on the "theoricae planetarum" in 1594–1595 and 1604–1605 (see ibid., pp. 119–120).

[4] For the Pisan *rotuli*, see the references reported *supra*, Chapter 2, n. 2.

[5] See EN, I, pp. 7–177. Antonio Favaro included these writings, which are part of Ms. Gal. 46, in the National Edition under the title "*Iuvenilia*," using a slightly different

various sources, offers insight into Galileo's early academic exposure. Although their exact timeline is debated, it is believed they were written before 1592.[6] In the context of our research, these writings provide a richer understanding of Galileo's perspectives in *De motu antiquiora*.

Secondly, both in the *Juvenilia* and in *De motu antiquiora*, Galileo seems to refer to the chapter division of the 1515 Venice edition of the *Almagest* translated from Arabic by Gerard of Cremona—a copy Galileo owned.[7] The 1528 Venice edition translated from Greek by George of Trebizond follows a slightly different chapter division. Another division is seen in the 1538 Greek *editio princeps* and in all subsequent translations based on it.[8] Although this suggests Galileo used the 1515 edition, it does not rule out his consultation of other versions of the *Almagest*. Actually, the chapter division of the first book in the 1515 and 1528 editions of the *Almagest* is almost identical. The following is a comparison between the chapter division of the first book of the *Almagest* from the Latin versions that Galileo might have read:

spelling. However, Galilean scholars usually prefer the classical Latin orthography of the word. The *Juvenilia* were fully translated into English by Wallace: see Wallace 1977.

[6] For a general overview of the issue regarding the date of composition of the *Juvenilia*, see Malara 2019, pp. 2–4, where relevant literature is also provided in n. 1.

[7] See Favaro 1886, p. 252.

[8] According to Toomer, Ptolemy divided the *Almagest* in books, but he did not use any chapter division (see Toomer 1984, p. 5). Apparently, chapter titles in Greek were later interpolations. Considering the heading of the chapter about the sphericity of the heaven in the Arabic versions of the *Almagest* by Ishaq (revised by Thabit) and al-Hajjaj, Langermann has shown that "the chapter title in Arabic does not reflect the Greek interpolation." See Langermann 2020, pp. 160–161. For a comparison between the chapter division and headings in the beginning of three versions of the first book of the *Almagest* (in Latin, published in 1515 and 1549, and in Greek, published in 1538), see Omodeo & Tupikova 2013, p. 240.

Tr. Gerard of Cremona (Venice 1515)	Tr. George of Trebizond (Venice 1528)	Tr. Erasmus Reinhold (Basel 1549)
I.1 De scientie Astronomie ad alias excellentia et finis eius utilitate	[I.1 Prohemium, Sive proloquium, Prologus.][9]	[Incipit][10]
I.2 De ordinibus modorum huius scientie	I.2 De ordine huius doctrinae	I.1 De ordine huius doctrinae
I.3 Quo scitur quod celum sit sphericum, et motus eius circularis	I.3 Quod sphaericum est, globique modo coelum convolvitur	I.2 Quod coelum sit sphaericum, et globi modo circumvolvatur
I.4 De eo quod indicat quod terra sit spherica	I.4 Quod terra quoque sphaerica sit ad sensum quantum ad universas partes	I.3 Quod terra sit sphaerica ad sensum secundum universas partes
I.5 De eo quod indicat quod terra sit in medio celi	I.5 Quod terra in medio coeli sita est	I.4 Quod terra in medio coeli sita sit
I.6 De eo quod indicat quod terra sit ut punctum apud celum	I.6 Quod terra quasi punctum est ad caelestia comparata	I.5 Quod terra velut punctum sit ad coelum collata
I.7 De eo quod indicat quod terra motum localem non habet	I.7 Quod terra nullo motu progressivo movetur	I.6 Quod terra non moveatur locali motu, seu mutatione loci
I.8 Quo declaratur quod primi motus qui sunt in celo sunt duo	I.8 Quod duplex in coelo primorum motuum differentia est	I.7 Quod primi motus in coelo sint duplices
I.9 De scientia quantitatis chordarum partium circuli	I.9 De quantitate rectarum linearum quae in circulo perducuntur	I.8 De particularium scientia
I.10 Quomodo tabule chordarum partium circuli fiant	[I.10 Tabulae arcuum et chordarum.]	I.9 De quantitate rectarum linearum in circulo
I.11 De positione arcuum et chordarum eorum in tabulis	I.11 De arcu qui est inter tropicos	I.10 De circumferentia inter tropicos
I.12 De arte instrumenti, quo scitur quantitas arcus qui est inter duo tropicos	I.12 Theoremata quae ad sphericas demonstrationes praemittuntur. Et ipsa figura sectoris sphaerica	I.11 Theoremata praemittenda sphaericis demonstrationibus
I.13 De scientia quantitatum arcuum qui sunt inter orbes equationis diei, et orbem medij signorum qui sunt declinationis	I.13 De arcubus qui sunt inter aequinoctialem atque obliquum circulum	I.12 De circumferentijs inter aequinoctialem et obliquum circulum
I.14 De scientia quantitatis arcuum equationis diei qui elevantur in sphera directa cum arcubus orbis signorum dati	I.14 De ascensionibus in recta sphaera	I.13 De ascensionibus in recta sphaera

For the sake of scrupulousness, these three Latin versions will be mentioned as needed. However, to avoid confusion, I will stick to the chapter division used by Galileo. Conveniently, the chapters of the *Almagest* explicitly referenced in *De motu antiquiora* share the same

[9] This list refers to the actual chapter division and titles running through the text of the first book of the 1528 edition. Omissions are integrated in square brackets, referring to the index page of the same edition.

[10] This Latin translation begins with an unentitled incipit.

numbering as those in the corresponding modern English translation by Gerald J. Toomer.

Thirdly, a brief overview of Galileo's earlier writings on motion is essential. Likely written between 1589 and 1592, *De motu antiquiora* contains various original works, including a dialogue between two characters, Alexander and Dominicus, three versions of a treatise, and miscellaneous notes. The central theme is the movement of heavy bodies.[11] In late sixteenth-century Pisa, this topic was discussed in "*disputationes circulares,*" debates including both university students and professors. The fascination with the motion of heavy bodies is evident in works by scholars like Buonamici, Mazzoni, and Girolamo Borro, who wrote *De motu gravium et levium* in 1575.[12]

CHALLENGING ARISTOTLE THROUGH MATHEMATICS: "THE DIVINE PTOLEMY AND THE MOST DIVINE ARCHIMEDES"

Although Galileo's primary subject as a professor was mathematics, he was deeply engaged in the discourse on natural philosophy.[13] He firmly believed that mathematics was the key to distinguishing truth from falsehood and comprehending "whatever is truly known among mortals."[14] In *De motu antiquiora*, the theoretical approach he employed to address certain issues concerning the motion of heavy bodies stood in clear

---

[11] For more information, see *supra*, Chapter 1.

[12] See Camerota & Helbing 2000.

[13] Galileo was not the sole professor of mathematics in Pisa with an interest in natural philosophy. Filippo Fantoni, who occupied the mathematics chair before Galileo, was also interested in the discourse regarding the motion of heavy and light bodies. For further details, refer to Schmitt 1972, esp. p. 260, n. 90; Schmitt 1978, esp. p. 61.

[14] In the dialogued version of *De motu antiquiora*, presumably the first written, Galileo let Dominicus express the following rhetorical question: "Let them be silent, certainly silent, those who think that philosophy can be attained without knowledge of divine mathematics. Who would ever deny that, guided by mathematics alone, it is possible to distinguish the true from the false, to sharpen the acumen of ingenuity with its help, and finally, under its guidance, to perceive and understand whatever is truly known among mortals?" (EN, I, p. 401: "Sileant profecto, sileant, qui philosophiam consequi posse autumant absque divinae mathematicae cognitione. Ecquis unquam negabit, hac sola duce verum a falso dignosci posse, huius auxilio ingenii acumen excitari, hac denique duce quicquid inter mortales vere scitur percipi et intelligi posse?").

opposition to Aristotelian views, particularly because it drew from mathematical tools not present in Aristotle's work. For example, Galileo refuted the idea that a continuously increasing speed could ever become infinite. He articulated this rejection as a criticism of Aristotle, using an analogy inspired by the concept of "asymptote" in Greek mathematics.[15] He also highlighted Aristotle's geometric limitations by comparing them to the mathematical prowess of Archimedes, whom he described as "superhuman."[16]

In *De motu antiquiora*, Galileo's respect for the mathematicians he studied under is evident. Ptolemy was clearly one of these influential figures, standing alongside others like Euclid, Apollonius, and Archimedes.[17] In the dialogic version of *De motu antiquiora*, Galileo praised Ptolemy as "divine." Along with Archimedes, whom Galileo termed as "most divine," Ptolemy was commended for his "most certain, clear, and subtle mathematical demonstrations." The character Dominicus expresses this sentiment after advancing six motion-related queries to Alexander[18]:

> I would greatly appreciate hearing your opinion on these matters and others related to them. For I know that on this subject, you will either

---

[15] See EN, I, pp. 328–331. For an accurate analysis of this passage, see Sisana 2023, pp. 219–233.

[16] See EN, I, p. 300, l. 18, where Archimedes is described as *superhumanus*, and pp. 302–304, where Galileo claims that "Aristotle was little versed in geometry" (*Aristotelem parum in geometria fuisse versatum*). He illustrates this by showing that, contrary to Archimedes, Aristotle wrongly believed that a straight line and a curve are not comparable.

[17] In a letter to Christoph Grienberger, Galileo mentions these four authorities in mathematics as those who the reader should know to appreciate the arguments of the *Sidereus Nuncius* (EN, XI, *Galileo to Grienberger* (1 September 1611), p. 201: "Io [...] ho sempre supposto di parlare a persone di qualche prattica nella geometria, le quali, esercitate in Euclide, in Archimede, in Apollonio, in Tolomeo et altri, sappino come nelle dimostrationi delle passioni de i solidi frequentissimamente si seghino con i piani, et sopra le loro settioni si formano le figure et le dimostrationi insieme"). Later, after 1632, Galileo mentions again these very same authorities in his annotations on Rocco's exercises (EN, VII, *Notes on Rocco's Ex. Ph.*, p. 744: "Ma passiamo pure a considerare quello che scrivete, Sig. Rocco mio, nelle 2 seguenti facciate: concetti composti di parole matematiche, ma tali che io, che ne fo professione e che ho inteso quello che scrivono Euclide, Archimede, Apollonio, Tolomeo ed altri molti celebri autori, non ne so trar costrutto alcuno").

[18] For Dominicus' queries, see EN, I, p. 308. See also Wallace's summary in Wallace 1984b, pp. 232–233.

say nothing or bring forth something new and closest to the truth. *Given that you are familiar with the most certain, clear, and subtle mathematical demonstrations, like those of the divine Ptolemy and the most divine Archimedes*, you cannot possibly agree with some of the more superficial arguments. And since the points I have raised are not far from mathematical considerations, I eagerly expect some insightful remarks from you.[19]

According to Dominicus, Alexander is "familiar" (*assuetus*) with the proofs of Archimedes and Ptolemy. Because of this familiarity, he is urged to share his views on the motion of heavy bodies, answering questions that are "not far from mathematical considerations" (*non longe a mathematicis considerationibus dist*[a]*nt*).

The character Alexander has been interpreted as a representation of Galileo, similar to Salviati in both *Dialogue* (1632) and *Two New Sciences* (1638).[20] If this interpretation holds, the aforementioned passage would testify to the deep familiarity Galileo possessed, or believed he possessed, with the works of Ptolemy and Archimedes as he set out to write *De motu antiquiora*. Regarding the implied reference to Ptolemy's work, it is plausible that Galileo was hinting at the proofs within the *Almagest*. The latter is indeed the only work of Ptolemy referred to in *De motu antiquiora*.

---

[19] "Tuam itaque sententiam de his et de similibus, quae ab istis pendent, audire graditissimum erit: scio enim te in hac materia aut nihil dicturum, aut aliquid novi et veritati ipsi propinquissimum in medium allaturum. *Cum enim certissimis, clarissimis atque subtilissimis mathematicis demonstrationibus sis assuetus, utpote divini Ptolemaei et divinissimi Archimedis*, crassioribus quibusdam rationibus nullo pacto assentiri potes: cumque haec, quae proposui, non longe a mathematicis considerationibus distent, abs te aliqui pulcri arrectis auribus expecto" (EN, I, p. 368, emphasis added).

[20] As Favaro argued in EN, I, p. 148: "Del 'Dominicus' nulla sappiamo; ma certamente 'Alexander' altri non è che Galileo stesso, perché in un certo luogo Alessandro parla della bilancetta come d'uno strumento da lui inventato (pag. 379), ed ancora servendosi di termini da Galileo stesso adoperati nella relativa scrittura." Among the scholars who accepted this interpretation, there are also Drabkin in Galilei 1969, p. 57, Fredette in Fredette 1969, p. 40, and Wallace in Wallace 1984b, p. 233.

Experiencing The Earth's Central Position in the World

Ptolemy's name also appears in the treatise version of *De motu antiquiora*.[21] At the beginning of the first version of the treatise, where "philosophers"—Aristotle included—are criticized for not clarifying why the Earth is at the center of the universe,[22] Galileo adds on the margin:

> Ptolemy, at the beginning of the seventh chapter of his *Construction* [viz. the *Almagest*], says it is vain to inquire why heavy objects are drawn to the center [of the universe], having demonstrated that the Earth, to which they are drawn, is in the center.[23]

The passage is subsequently incorporated into the main text and is nearly identical in the other two versions of the treatise. For convenience, I will reference only what is generally considered to be the last version written:

> In this [geocentric] order, therefore, nature has arranged bodies, so that, specifically, the heavier ones remain closer to the center. *This arrangement is constantly affirmed by our experience.* However, one might question why the wise nature preserved such an order in distributing places and not the reverse. The philosophers offer no other reason, as far as I have read, for this distribution, except that everything had to be ordered in some way, and the Supreme Wisdom chose this order. Aristotle seems to suggest something similar in *Physics* VIII, text 32, when, asking why heavy and light objects move to their proper places, he supposes that the reason is because they are naturally inclined to move somewhere – the light objects

---

[21] For a brief description of the occurrences of Ptolemy's name in *De motu antiquiora*, see Sisana 2023, pp. 234–236.

[22] Note that Sacrobosco, in his *De Sphaera*, referred to the authority of "Ptolemaeu[s] et omnes philosoph[i]" (see Thorndike 1949, pp. 84 (Latin) and 122 (English translation)). See also Omodeo & Tupikova 2016, p. 172.

[23] "Ptolemaeus, in principio 7 cap<ituli> p<rim>i suae Constructionis, ait frustra inquiri cur gravia ad medium ferantur; cum demonstrasset Terram, ad quam feruntur, in medio esse" (EN, I, p. 252).

upward and the heavy ones downward. However, Ptolemy, at the beginning of the seventh chapter of the first book of his *Great Construction*, says it is vain to inquire why heavy objects are drawn to the center [of the universe], having demonstrated that the Earth, to which they are drawn, is in the center.[24]

In *De motu antiquiora*, Galileo openly advocates for a geocentric system. He seems to be convinced of this based on an unspecified "experience" which consistently reveals the Earth's central position in the world (*"...continua nobis declarat experientia"*). In the alternate two versions of the treatise and the dialogic rendition, the Latin term *"sensus"* is used in place of "experience," denoting what can be perceived by the senses.[25] In each of these versions, sensory evidence reinforces the claim that the Earth is in the center of the world or universe.[26]

Understanding which experience Galileo alludes to is not straightforward. Nevertheless, considering the most common and traditional arguments of the time, it is highly probable that he had in mind the demonstrations of Ptolemy. In *De caelo*, Aristotle addresses the issue of geocentrism with only one reason based on appearances. He refrains from delving into details, simply acknowledging that mathematicians have demonstrated, in the realm of astronomy, that what is observed in the

---

[24] "In hunc, itaque, ordinem a natura distributa fuisse corpora, ut, scilicet, quae graviora essent, centro propinquiora manerent, *continua nobis declarat experientia*: verum in dubium revocari potest, cur talem ordinem in distribuendis locis, non autem praeposterum, prudens natura servaverit. Huius distributionis non alia, quod legerim, a philosophis adfertur causa, nisi quod in aliquem erant ordinem cunta disponenda, in hunc autem Summae Prudentiae distribuere placuerit. Simile quiddam Aristoteles, 8 Phys<icorum> 32, adferre videtur, dum, quaerens cur gravia et levia ad propria moveantur loca, subdit, causam esse quia habent a natura ut sint apta ferri aliquo, et hoc leve quidem sursum, grave autem deorsum. Ptolemaeus autem, in principio 7i cap<ituli> p<rim>i libri suae Magnae Constructionis, inquit frustra inquiri cur gravia ad medium ferantur; cum demonstrasset Terram, ad quam feruntur, in medio esse" (EN, I, 344–345, emphasis added).

[25] See EN, I, pp. 374, 252, 342.

[26] On the alleged heliocentric positions of Galileo in *De motu antiquiora*, see Chapter 1. Here, I simply recall that, even assuming Galileo was already a Copernican when composing *De motu antiquiora*, his exposition of geocentrism remains distinctive and noteworthy, as will become evident later on.

heavens presupposes the immobility of the Earth at the center.[27] Ptolemy, on the other hand, provides a detailed account of the astronomical observations that confirm, via *reductio ad absurdum*, the central position of the Earth in the world. Given Galileo's pronounced polemical stance against Aristotle, it may be appropriate to confine exploration in this context to the hypothesis that, when Galileo penned *De motu antiquiora*, he was influenced by the arguments in the first book of the *Almagest*. In this book, Ptolemy, indeed, utilizes astronomical data and observations to argue for the Earth's central position in the universe.

At the start of *Alm.* I.5, Galileo would have encountered the following statement[28]:

Tr. Gerard of Cremona (Venice 1515)	Tr. George of Trebizond (Venice 1528)	Tr. Erasmus Reinhold (Basel 1549)
[...] if we desire to know the place of the Earth in the matters we will discuss, such knowledge will not be fulfilled *by what appears to us from the Earth, just as we see and find out*, unless we affirm that the Earth's place is the middle of the sky, just like the center of a sphere[29]	[...] if someone subsequently wants to speak more precisely about the position of the Earth, he will undoubtedly understand that *the things appearing around it* occur only if he places it at the middle of the sky as if at the center of a sphere[30]	[...] if someone subsequently seeks the position [of the Earth], he will realize that *what we have said about the apparent positions of stars* can only be the case if we place the Earth at the middle of the sky like a center[31]

[27] See *De caelo*, II, 296 a 34–296 b 6, 297 a 2–6.

[28] See the English translation from the Greek in Toomer 1984, p. 41: "[...] if one next considers the position of the earth, one will find that the phenomena associated with it could take place only if we assume that it is in the middle of the heavens, like the centre of a sphere" (emphasis added).

[29] "[...] si scire cupierimus locum terre in his que narrabimus, non complebitur eius scientia *per hoc quod apparet nobis in ea sicut videmus et reperimus* nisi cum affirmaverimus locum eius medium celi sicut centrum in sphera tenere" (Ptolemy 1515, f. 3v, emphasis added).

[30] "[...] siquis deinceps de situ terrae certius dicere velit sic profecto *quae iuxta ipsa apparent*, accidere solummodo intelliget si tam in medio coeli quasi sphaerae centrum posuerit" (Ptolemy 1528, f. 2v, emphasis added).

[31] "[...] si quis deinceps locum inquirat, deprehendet *ea, quae de adparentijs stellarum diximus*, ita tantum posse accidere, si collocemus terram in medio coeli tanquam centrum" (Ptolemy 1549, f. 53r, emphasis added).

Ptolemy argues that the notion of an Earth not centrally located (relative to the sphere of the fixed stars) contradicts our observations, such as those concerning equinoxes and solstices. It also does not match our perceptions regarding our distance from the stars, the number of zodiacal constellations visible above the horizon, or the correlation between the positions of the gnomon's shadow at dawn and sunset on equinoctial days. This chapter also briefly touches on lunar eclipses.[32]

In the following chapter (*Alm.* I.6), Ptolemy maintains that, based on our senses (variously translated as *"secundum sensum," "quantum ad sensum pertinet," "quod ad sensum attinet"*), the Earth can be effectively idealized as a point in the center of the universe compared to the vast distance separating it from the fixed stars. This is underscored by three key observations: (1) from any location on Earth and at any time, we perceive a consistent distance to the fixed stars; (2) no matter our location on Earth, gnomons act as though they are located at the center of the universe, and armillary spheres appear as if their center is equidistant from the fixed star sphere; (3) the horizon always seems to split the sphere of fixed stars in half, creating two hemispheres.[33]

These arguments were widely recognized during Galileo's time and were explored in more accessible writings, like commentaries on the

---

[32] See Pedersen 2011, pp. 41–42. Aristotle also addresses this in *De caelo*, but with a different objective in mind: to prove the Earth's spherical shape (cf. *De caelo*, II, 14, 297 b 23–30).

[33] For a concise yet clear and precise overview of Ptolemy's arguments in *Alm.* I.5–6, see Pedersen 2011, pp. 39–43; Taub 1993, pp. 72–74; Evans 1998, pp. 76–78, esp. Figures 2.3–6. On *Alm.* I.5, see also Omodeo & Tupikova 2016, pp. 161–165.

Sphere and various cosmography treatises.[34] Within these works, Ptolemy's arguments were often integrated with insights from other thinkers, including Aristotle, Averroes, al-Farghani, and Regiomontanus.

In the *Trattato della sfera overo cosmografia*, probably written in Padua and primarily used for private instruction, Galileo discussed only Ptolemy's arguments.[35] A compilation more reminiscent of *Sphere* commentaries, frequently referencing other scholars, appears in the *Juvenilia*.[36] This text also cites "mathematicians" who highlight discrepancies between the phenomena we would expect to observe if the Earth were off-center and the actual "daily experience" we encounter.[37]

So, it is likely that when Galileo was in Pisa, writing *De motu antiquiora*, he was already acquainted with these arguments. Notably, while in Pisa, he seemed to be convinced by these viewpoints. However, by the time he penned his *Cosmografia*—assumed to be after 1597, the

---

[34] Edward Grant believed that these arguments were original to Clavius. For instance, see Grant 1984, p. 31, where the "three alternatives" (n. 111) he attributes to Clavius are those of *Alm.* I.5. This is reiterated in a more recent article (see Grant 2003, p. 129). Evidently, Grant did not notice that before presenting such alternatives, in the same 1593 edition he read, Clavius says: "prima [ratio] desumpta ex Ptolem. Dict. 1 cap. 5 sit haec." It is true that the three alternatives were not detailed in Sacrobosco's *Sphere* (see Thorndike 1949, pp. 84 in Latin, and 122 in English translation), but they were certainly not introduced for the first time by Clavius. Alessandro Piccolomini, just to give an example, recalls the same arguments by attributing them to Ptolemy, and one of these, also to Averroes (see Piccolomini 1540, ff. 11r–v; see also the more elaborated version in Piccolomini 1561, ff. 12v–14v). Unfortunately, Grant did not recognize many of the original sources of the arguments employed by the authors he considered. The belief that if the Earth were outside the center of the world, "the same stars ought to appear larger when nearer the Earth and smaller when farther away" (Grant 1984, p. 31) is clearly stated by Sacrobosco in the *Sphere* (refer again to Thorndike 1949, pp. 84, 122). Additionally, the opinion that lunar eclipses confirmed that earth and water formed a spherical globe was not only a "popular astronomical argument" (Grant 1984, p. 28) but came directly from Aristotle's *De caelo*, II, 14, 297 b 23–30. The other argument that "relied on the experiences of sailors" is also present in *Alm.* I.4 and Sacrobosco's *Sphere* (see Thorndike 1949, pp. 83, 121). Also, Grant's citation of Aversa's arguments against Copernicus simply restates those put forth by Ptolemy in *Alm.* I.5 (see Grant 1984, p. 32).

[35] See EN, II, pp. 220–221.

[36] See EN, I, pp. 48–50.

[37] "[…] quae omnia probant mathematici quotidianae experientiae repugnantia" (EN, I, p. 49). Cf. *De caelo*, II, 297 a 2–6.

year Galileo disclosed to Mazzoni and Kepler his Copernican leanings[38]—he seemed to have changed his opinion.

GEOCENTRIC AND GEOSTATIC REASONS

At this point, it might be worthwhile to recall that in *De Caelo*, the centrality of the Earth in the world and its immobility are treated as a unified issue.[39] In the *Almagest*, however, the arguments in favor of the Earth's central position are laid out in the first place. It is only later that Ptolemy argues that the Earth's motion around its axis is illogical for physical reasons.[40] This distinction between the question of placement and that of the motion of the Earth was commonly accepted during Galileo's time. It is not surprising, considering the same differentiation was already present in the highly successful university manual that was Sacrobosco's *Sphere*.[41]

As previously observed, in Galileo's *De motu antiquiora*, there appears to be a firm belief that the Earth occupies a central position in the universe. This conviction is rooted in sensory perception and experience, likely referencing observations reported by Ptolemy in *Alm.* I.5–6, which establish the central position of the Earth. It seems that the question of the diurnal motion of the Earth is consequently set aside, not explicitly considered by Galileo in *De motu antiquiora*. This, in turn, leaves room for the possibility that during that period, Galileo had already theoretically entertained the notion that the Earth, at the very least, might move around its axis. Indeed, a few passages of *De motu antiquiora* could be interpreted in such a manner.[42]

---

[38] According to Drake, "it was probably for the purposes of his private teaching in 1586–87 that Galileo originally composed a manuscript 'Treatise on the Sphere, or Cosmography'" (Drake 1978, p. 12). However, this is not certain. As for the letter to Mazzoni and Kepler, see EN, II, *Galileo to Mazzoni* (30 May 1597), pp. 197–202; EN, X, *Galileo to Kepler* (4 August 1597), pp. 67–68 (partially reported *supra*, Chapter 1, n. 45).

[39] See *De caelo*, II, 14, 296 a 24–297 a 6.

[40] See Pedersen 2011, pp. 39–44; Toomer 1984, pp. 41–45.

[41] See Thorndike 1949, pp. 84–85 (Latin), 122 (English translation).

[42] See EN, I, pp. 304–305. See also p. 373 (dialogued version). For insights into the interpretations of these passages in terms of geokinetic or even Copernican perspectives, see *supra*, Chapter 1.

However, despite the distinct nature of the arguments for the Earth's centrality and immobility at that time, Galileo, in a later and more advanced stage, contends that the soundness of Ptolemy's arguments supporting the centrality of the Earth relied on "another assumption," specifically, the immobility of the Earth. This viewpoint is articulated in his letter to Francesco Ingoli dated 1624.[43]

In 1616, Ingoli authored *De situ et quiete Terrae contra Copernici systema disputatio*, wherein he critiqued the Copernican system using arguments categorized as theological, mathematical, and physical.[44] As indicated by the complete title of the *Disputatio*, Ingoli first presents criticisms against the positions assigned by Copernicus to the Earth and the Sun. Subsequently, he articulates his own reasons against the motion of the Earth and the stillness of the Sun.

Ingoli derives the third of the mathematical arguments against the Copernican position of the Earth from Ptolemy. Initially, he appeals to some of the arguments presented in *Alm.* I.5–6. Then, to support them, he draws upon data taken from Kepler (though not explicitly cited), Brahe, and Magini.[45]

Galileo's response to this argument is extensive and detailed. It may be useful, however, to consider its beginning to better grasp how, in Galileo's view, certain evidences supporting geocentrism could coherently be regarded as true and consequently become objects of experience for those unconsciously assuming the Earth's immobility.

Firstly, Galileo reminds that concerning the issue of the position of the Earth, some reasons are true while others are false;

> and among the false ones, sometimes there may be one that bears some aspect of truth, in comparison to others that, upon any moderate scrutiny, reveal themselves for what they are – false and irrelevant to the matter at hand [...]. Of those that, at first glance, may have some aspect of truth, there is this one that you take from Ptolemy, just as there are others presented by him in his *Almagest*. These *not only appear true but, I will*

---

[43] See EN, VI, pp. 509–561. On this letter, see Bucciantini 1995, pp. 149–174.

[44] The *Disputatio* is fully transcribed in EN, V, pp. 403–412. On the *Disputatio*, see Bucciantini 1995, pp. 88–97.

[45] See EN, VI, pp. 405–406. Bucciantini highlighted the importance of the authority of Tycho in Ingoli's arguments (see Bucciantini 1995, pp. 90–97). For Ingoli's reference to Kepler's *De stella nova in pede Serpentarii*, see Galilei 2009, p. 171 n. 11.

say, they are also conclusive within the entire Ptolemaic position, yet they prove to be inconclusive within the entire Copernican system.[46]

It is well-known that Galileo dismissed Brahe's system as "null" (*nullo*) while considering both Ptolemy's and Copernicus' as "complete" (*interi*).[47] In this passage, one might glean one of the reasons behind Galileo's stance. The Ptolemaic and Copernican systems, viewed comprehensively, offer "conclusive" propositions, meaning they align with the entire theoretical structure from which they stem. Some propositions, true in one system, prove false in the other. Certainly, truth is not contingent on the favored reference system but must be ascertained in light of the "whole university of nature" (*università della natura*).[48] However, certain propositions,

> sometimes, when they are attached to another false proposition, can be conclusive with that assumption. An example of this will be the discussion we currently have at hand. You [Ingoli] say, following Ptolemy: If the Earth were not at the center of the stellar sphere, we could not always see half of that sphere, but we do see it; therefore, etc. [...] The argument is elegant and worthy of Ptolemy, and *coupled with another assumption, it necessarily concludes*. However, if that assumption is denied, the argument becomes null.[49]

[46] "[...] e tra le false alcuna tal volta ve ne può essere che abbia qualche sembianza di verità, in comparazione di altre che ad ogni mediocre discorso si rappresentano quali elle sono, cioè false e fuori del caso [...]. Di quelle che nel primo aspetto abbiano qualche sembianza di verità, ne è questa che voi prendete da Tolomeo, sì come sono anco altre prodotte dal medesimo nel suo Almagesto, *le quali non solamente hanno aspetto di vero, ma dirò che sono anco concludenti nell'intera posizione Tolemaica*, ma bene nulla concludenti nell'intero sistema Copernicano" (EN, VI, p. 526, emphasis added).

[47] See, for instance, EN, VI, *The Assayer*, pp. 232–233. Additional references can be found in Malara 2023, p. 468 n. 16.

[48] "Adunque, direte voi, possono le medesime proposizioni concludere e non concludere, ad arbitrio altrui? Signor no, prese assolutamente ed in tutta l'università della natura" (EN, VI, p. 526).

[49] "[...] attaccate tal volta ad un'altra proposizione falsa, possono esser, con quella supposizione, concludenti: esempio di che vi sarà il discorso che ora aviamo alle mani. Voi dite con Tolomeo: Se la Terra non fusse nel centro della sfera stellata, noi non potremmo veder sempre la metà di esse sfera; ma noi la veggiamo; adunque etc. [...] Il discorso è bello e degno di Tolomeo, ed *accoppiato con un'altra supposizione, conclude necessariamente*; ma negata quella, l'argomento resta nullo" (EN, VI, pp. 526–527, emphasis added).

Galileo hints at the immobility of the Earth, as he later clarifies when asserting that

> it is true that, with two fixed stars alternately rising and setting near all horizons [*presso tutti gli orizonti*], one must necessarily assert that the Earth is in the center of the sphere of the [fixed] stars. However, this is contingent upon the Earth being immobile and the rising and setting arising from the motion and rotation of the sphere of the stars. But if we (as Copernicus does) keep this sphere stationary and revolve the terrestrial globe within itself, placing it wherever you like, the same will always happen with the two fixed stars, namely, their alternate rising and setting.[50]

Then, a geometric proof follows.[51] Here, though, it is particularly interesting to linger on how the mature and openly Copernican Galileo views Ptolemy's arguments regarding the centrality of the Earth. These are not simply dismissed as false. They are classified among the arguments that "bear some aspect of truth." Furthermore, they "necessarily conclude" when a false premise is accepted—specifically, the Earth's immobility. Unless one grasps the falsehood of this premise, the arguments in *Alm.* I.5–6 retain their persuasiveness. As long as the motion of the Earth is not taken into serious consideration, one can indeed experience the Earth's centrality in the world.

The mature Galileo, thus, provides us with a possible explanation for why he, in his earlier writings on motion, believed that one could rely on sensory experience to establish the centrality of the Earth in the world. It may be that, by embracing the foundational assumption of the Ptolemaic system, specifically the Earth's immobility, the proposition of the Earth's centrality appeared inherently conclusive to him.

---

[50] "[...] è vero che nascendo e tramontando alternamente appresso tutti gli orizonti due stelle fisse, bisogna per necessità dire, la Terra esser nel mezzo della sfera stellata, tuttavolta però che la Terra stia immobile e ch'il nascere e tramontare derivi dal moto e conversione della sfera stellata: ma se noi (come fa il Copernico) faremo star ferma la sfera e rivolgere in se stesso il globo terrestre, ponetelo pur poi dove più vi piace, che sempre avverrà delle due stelle fisse quello che si è detto, cioè il nascere e tramontare alternamente" (EN, VI, p. 527).

[51] See EN, VI, pp. 527–528. Note that this geometric proof differs from the one Galileo sent to Mazzoni in 1597, where Mazzoni's premises, including the Earth's immobility, are assumed but not conceded (see EN, II, pp. 197–202; mentioned *supra*, Chapter 2).

WHY IS THE EARTH IN THE CENTER OF THE WORLD? ARISTOTELIAN AND NON-ARISTOTELIAN INTERPRETATIONS OF PTOLEMY

In *De motu antiquiora*, Galileo not only addresses the Ptolemaic geocentric system as if it were proven by the senses but also delves deeper than Ptolemy, exploring the "why" (*cur*) behind nature's placement of heavy objects at the world's center. Ptolemy, at the beginning of *Alm.* I.7, which Galileo refers to, posits that once one demonstrates the Earth's central position, questioning the reason becomes unnecessary[52]:

Tr. Gerard of Cremona (Venice 1515)	Tr. George of Trebizond (Venice 1528)	Tr. Erasmus Reinhold (Basel 1549)
Therefore, I have considered that investigating the causes of motion toward the center is superfluous and vain, given what has already been shown concerning what is observed—that the Earth is in the central location of the world, and that all heavy things from every direction hastily tend toward the Earth[53]	Thus, if anyone seeks the reasons for the motion of heavy things toward the center, it seems to me he is doing so in vain, since it is clearly manifest in reality that the Earth occupies the center of the world, and all weighty things are drawn to Earth[54]	Therefore, it is unnecessary to seek the reason why heavy objects move toward the center, since it is manifest from appearances that the Earth holds the central position, and all weighty things are drawn to the Earth[55]

[52] English translation in Toomer 1984, p. 43: "Hence I think it is idle to seek for causes for the motion of objects towards the centre, once it has been so clearly established from the actual phenomena that the earth occupies the middle place in the universe, and that all heavy objects are carried towards the earth."

[53] "Quapropter estimavi quod investigare causas motus qui est ad medium superfluum est et vanum post illud quod iam semel ostensum est de hoc quod videtur quod terra sit in loco medii mundi et quod gravia omnia undique ad ipsam festinanter tendant" (Ptolemy 1515, f. 4r).

[54] "Qua re siquis causas motus gravium ad medium quaerat, frustra mihi facere videtur, cum reipsa manifestissimum sit et terram medium mundi loco possidere, ponderosaque omnia fieri ad ipsam" (Ptolemy 1528, f. 3r).

[55] "Quare supervacaneum est causam quaerere, cur gravia moveantur ad medium, cum ex adparentibus manifestum sit, quod et terra medium loco teneat, et ponderosa omnia ad ipsa ferantur" (Ptolemy 1549, f. 63v).

It is crucial to highlight that neither Regiomontanus' (and Peurbach's) epitome mentions Ptolemy's stance on the needless inquiry into why heavy objects move toward the center of the universe. This standpoint is not mentioned in Barozzi's *Cosmographia*, Clavius' commentary on the *Sphere*, or Alessandro Piccolomini's *De la Sfera del mondo divisa in quattro libri* (1540)—all of which were in Galileo's library.[56] This particular point is not cited in the *Juvenilia* as well. Thus, it seems plausible that Galileo acquired this insight directly from the *Almagest*, referencing at least one of the existing Latin translations.

For clearer context on the *Almagest* segment Galileo references, it is important to note that some texts from that period tended to merge Aristotle's views on gravity with Ptolemy's.[57] For example, commenting on *Alm.* I.7, Schreckenfuchs remarked that

> since heavy bodies naturally move downward, and the Earth is the heaviest of all, its motion will only be *toward the center of the universe, to which it naturally tends*, much like iron to a magnet. Therefore, it should not be believed that it moves in a straight motion.[58]

---

[56] See Favaro 1886, pp. 249–250, 251, 260.

[57] During the Renaissance, Ptolemy was interpreted differently. The prologue to the Latin translation from Arabic depicts Ptolemy as a spiritual and moral guide, with his work, the *Almagest*, helping to deepen one's understanding of God (see Burnett 2010; Bianchi 2022, p. 302). Andrea of Trebizond, in his preface, suggested that Ptolemy came from the royal Ptolemaic dynasty, although by his time the dynasty had lost its power over Egypt. Despite this, Ptolemy was viewed as having the spirit and intellect of a monarch, remaining unaffected by worldly temptations. He devoted himself to scholarly pursuits beneficial to society, and with great skill (*"haud obscuris facultatibus"*), studied in the renowned library of Alexandria. Here, "totum se ad litteras contulit, atque in primis in philosophia praeclara, humanae societatis parente, Autore illo suo naturae interprete Aristotele contenta, deinde in mathematicis disciplinis (quibus vagantia Coelo sidera cognoscuntur) et quibus succurrendum videbat." His intellect was so sharp and nuanced that he could directly discern the truths of the natural world from nature itself (*"e naturae sinu"*). On Andrea of Trebizond's preface, which deserves further attention, see Fuiano 1967, esp. p. 19, n. 45, where the passages just quoted from Guarico's edition is collated with Ms. Vat. Lat. 2054, showing that Guarico's text sometimes differs in meaning from the original. For a contemporary understanding of Ptolemy's contribution to philosophy, see Feke 2018.

[58] "[…] cum gravia corpora naturaliter moveantur deorsum, et Terra sit gravissima, erit motus eius tantum *ad centrum universi, ad quod naturaliter tendit*, et non secus ac ferrum ad magnetem, ideo non credendum est eam moveri recto motu" (Ptolemy 1551, f. b3r, emphasis added).

Nature compels heavy bodies to move downward, specifically toward the center of the universe. Given that the Earth is the heaviest body, its natural motion is toward this central point. Yet, the Earth does not actually move in this way, as it already is in the center. This understanding mirrors that of Aristotle in *De caelo*. Specifically, in the second book, he clarifies that heavy objects move naturally toward the center of the universe, not necessarily toward the Earth.[59] This distinction is further emphasized in the fourth book with a counterfactual reasoning: if the Earth were positioned where the Moon is, objects would not move toward the Earth but would instead be drawn to the universe's center.[60] This confirms the Earth's central position in the world, with the Earth being the totality of heavy things. Schreckenfuchs essentially echoes this doctrine, introducing a fascinating analogy that likens the Earth's natural motion toward the universe's center to the pull of a magnet on iron. In his interpretation, Aristotle's and Ptolemy's perspectives on gravity align seamlessly.

A similar confluence of thought may be present in Reinhold's commentary on *Alm.* I.7:

---

[59] See *De caelo*, II, 14, 296 b 6–21, esp. 296 b 16–18.

[60] See *De caelo*, IV, 3, 310 b 3–5. The same counterfactual example is taken up by Simplicius in his commentary on the fourth book of *De caelo*, but to argue the opposite, that is, if the Earth were where the Moon is now, all earthly bodies would spontaneously return to the Earth. This passage has been overlooked by many scholars, especially those interested in Copernicus' sources on gravity (see Malara 2020). Aristotle's counterfactual example is also taken up by Alessandro Piccolomini. After recalling Ptolemy's arguments to confirm the central position of the Earth, Piccolomini concludes as follows: "Mà la più forte ragione, che prova la Terra essere in mezo del Mondo, è quella che si tra' dà Aristotile, ne i suoi libri del Cielo, dove prova che quanto un corpo è più grave, tanto più cerca di accostarsi al basso, verso il Centro del Mondo, mà la Terra essendo gravissima, è necessario che nel Centro stesso si posi; altrimenti s'ella fusse fuor del Centro, bisognaria che nel Centro fusse alcun corpo men grave di quella; Il che è falsissimo" (Piccolomini 1540, f. 11v; Piccolomini 1561, f. 14r). In the 1561 edition, the reference to Aristotle's counterfactual argument is more evident. It is used to show that the Earth is fixed and stable in the center of the world: "La onde la Terra tutta insieme, nonmeno che ciascheduna delle parti sue, ha in se gravezza, per la quale ogni volta che fosse fuora di quel Centro, à quello si muoverebbe di maniera che se per imaginatione noi ponessimo che la Terra tutta insieme fosse nel concavo del Cielo Lunare, quel medesimo impeto che nel Centro la tiene immobile, al medesimo Centro in quel caso la condurrebbe" (Piccolomini 1561, f. 15v). Aristotle's argument was also used by Buonamici (see Buonamici 1591, p. 458G-H).

Ptolemy says that there is no need to ask why heavy things move toward the center, as universally nature is so constituted that, *by their own and native impulse, things which are akin in nature desire the same place.* But Earth and other heavy things are akin in nature. And from other phenomena, it is clear that the Earth holds the central place. Therefore, other heavy things also move to the center [of the world], and having found their place there, they rest by themselves. *Indeed, says Ptolemy, this very identical fall of heavy or weighty things clearly proves that the Earth remains unmoved at the center.*[61]

Reinhold maintains that bodies sharing a similar nature, such as the Earth and heavy objects, inherently gravitate to the same location.[62] Given that Ptolemy has shown that the Earth lies in the center of the universe, other heavy bodies must naturally be drawn to this same center.

However, Reinhold also argues that Ptolemy distinctly conveys that heavy bodies are pulled toward the center of the universe, not specifically the Earth's center. So Reinhold deduces that, in Ptolemy's view, the Earth's immovable central position in the universe becomes evident when observing the descent of heavy objects, as they consistently move toward the world's center and do so in a path perpendicular to the Earth's surface.

Similar to Schreckenfuchs' interpretation, Reinhold's perspective can be perceived as a somewhat interpretive take on the *Almagest*, a viewpoint he seems to have adopted from Theon's commentary, which he explicitly cites.[63] This interpretation appears influenced by an argument Aristotle

---

[61] "Nihil opus est, inquit Ptolemaus, quaerere cur gravia moveantur ad medium. Quia universaliter sic condita est natura, *ut proprio et nativo impetu, ea quae sunt cognatae naturae, appetant eundem locum.* Sunt autem cognatae naturae Terra, ac caetera gravia. Estque ex alijs phaenomenis manifestum, quod Terram medium locum teneat. Ergo et caetera gravia ad medium feruntur, et sede ibi nacta, per se quiescunt. *Imo, inquit Ptolemaeus, hic ipse aequabilis casus gravium seu ponderosum aperte convincit Terram immotam quiescere in medio*" (Ptolemy 1549, f. 64r, emphasis added).

[62] On the influence of Copernicus in this interpretation, refer to Omodeo & Tupikova 2013, p. 251. However, note that the notions of "affinity" and "desire" are also found in Simplicius' commentary on *De caelo*. Also, while Reinhold claims that bodies of similar nature are drawn to the same place (*"proprio et nativo impetu, ea quae sunt cognatae naturae, appet[u]nt eundem locum"*), Copernicus posits that such bodies inherently tend to return to the entirety they belong to (see Copernicus 2015, II, p. 32, ll. 12–15).

[63] Theon's commentary in Greek is often quoted by Reinhold. See, for instance, Ptolemy 1549, ff. 74r, 98r, 103v, 105r, 106r, 107r, 116v.

puts forth in the second book of *De caelo*. Here, as recalled above, Aristotle first posits that heavy objects naturally move toward the center of the universe, and secondly, he notes they also converge toward the center of the Earth. Combining these observations, he deduces that the Earth must be at the center of the universe.[64]

However, Ptolemy's intent is not to validate the Earth's centrality based on the movement of heavy objects. Instead, he affirms that heavy objects move toward the center of the universe only after showing that they move toward an Earth that, for both geometric and astronomical reasons, occupies the center of the universe. This is further illuminated when considering Reinhold's translation of Ptolemy's text:

There is also this readily available argument to confirm this view [i.e., heavy things move toward the center of the world], that since the Earth is spherical and in the center [of the world], as we said, the inclinations and natural movements of all heavy bodies always and everywhere occur at right angles to that immovable plane, which is understood to touch the outermost surface of the Earth at the point where heavy things land.[65]

and immediately after, Ptolemy adds:

Given that this is the case, it is clear that heavy objects would seek *the very center of the Earth* if they were not held up by the Earth's surface. This is because a straight line aiming towards the center also forms right angles with that plane which touches the Earth's sphere at the point where the same line intersects the Earth's curvature.[66]

---

[64] See *De caelo*, II, 14, 296 b 6–22, esp. 296 b 18–21, where Aristotle employs the observation that bodies fall to the ground at equal angles as evidence supporting the notion that they also gravitate toward the center of the Earth, other than the center of the world. Thus, he concludes that the Earth is immobile in the center of the world (296 b 21–22).

[65] "Est et hoc argumentum in promptu, ad hanc sententiam confirmandam [i.e., gravia moventur ad medium mundi], quod cum Terra sit sphaerica, et in medio [mundi], ut diximus, gravium corporum omnium inclinationes et proprij motus semper et ubique fiunt secundum rectos angulos ei plano immoto, quod in loco incidentiae gravium extremam Terrae superficiem tangere intelligitur" (Ptolemy 1549, ff. 63v–64r).

[66] "Quod cum ita sit, manifestum est gravia petitura esse *ipsum Terrae centrum* si non a superficie Terrae sustinerentur, quia et recta linea ad centrum tendens ad rectos existit angulos ei plano, quod globum Terrae tangit in puncto, quo linea eadem secat Terrae convexitatem" (Ptolemy 1549, f. 64r, emphasis added).

It is evident that Ptolemy has no intention of using the trajectory of fall (always perpendicular to the tangent of the Earth's surface) to demonstrate that the Earth is at the center of the world. Yet, as previously mentioned, this is precisely how Theon interpreted his words, attempting to reconcile, even terminologically, Ptolemy's view with that of Aristotle. In Della Porta's translation, it reads:

Furthermore, from these points it becomes evident that the Earth itself remains in the center. *For, since it has a nature to be borne downward – as if the middle of a sphere is downward, just as Aristotle also calls the movement toward the center a movement "downward" –, when it reaches its natural place, it stays there.* Therefore, [Ptolemy says] "it seems to me redundant for someone to inquire about the reasons for the Earth's movement to the center, once it has been clearly established that the Earth occupies the central place." From which it follows that, since heavy objects naturally move downward, so they are indeed clearly seen to be borne toward the Earth, *thus, the Earth is in the center.*[67]

The last sentence seems to be a summary of Aristotle's argument in *De caelo* 296 b 6–21. Theon purposefully narrows the distinction Ptolemy seems to make from Aristotle's doctrine of gravity. Ptolemy suggests that terms like "above" and "below" are not applicable to the universe. In this respect, he seems to lean more toward Plato's insights in the *Timaeus*

---

[67] "Adhuc etiam ex his manifestum fit ipsam [i.e., Terram] in medio loco manere, *cum enim naturam ipsa habeat deorsum ferri, et tanquam in sphaera medio infra existente, quemadmodum etiam Aristoteles lationem ad medium deorsum vocat, haec cum ad proprium pervenit locum, in ipso manet.* Quare mihi quidem videtur superflue, quis et causas lationis Terrae ad medium inquireret, cum semel manifestum factum sit, quod Terra medium locum habet. Ex quo, quod et gravia habere naturam ut deorsum ferantur, ita enim manifeste ad Terram ferri videntur, *Terra igitur in medio est*" (Ptolemy & Theon 1605, p. 59; Della Porta 2000, p. 86, ll. 15–18, emphasis added).

than Aristotle's beliefs.[68] Similarly, in *De motu antiquiora*, Galileo emphasizes that we cannot label positions in the universe definitively, noting that "if we were to say, 'Heavy is that which remains below, and light that which remains above', we would not define it well, since 'above' and 'below' are distinguished not in reality, but merely with respect to something [*ratione*]."[69] However, in Theon's view, these terminological differences hold little significance. He finds it more crucial to highlight the convergence between Ptolemy and Aristotle. As a result, he attributes to Ptolemy an argument rooted in Aristotle's doctrine of natural motion.

It is intriguing that, more recently, Liba Chaia Taub has offered a similar interpretation. Referring to the aforementioned passage from *Alm.* I.7, Taub has argued that,

> so far, Ptolemy's principal argument against the Earth's motion from place to place was based on the idea that *the Earth must be at the center of the universe for largely physical reasons, because of the motion of heavy bodies.*[70]

However, Ptolemy does not appear to readily accept that heavy objects naturally move toward the center of the universe. Instead, he sees this as a proposition that requires demonstration. His argument for the Earth's centrality is not grounded in the movement of heavy objects, as Taub

---

[68] See Toomer 1984, p. 44, and Taub's exposition in Taub 1993, pp. 90–96: "Ptolemy's own discussion of direction terminology shares with those of both Plato and Aristotle an interest in accounting for common usage. However, Ptolemy clearly rejected Aristotle's opinion that directionality exists, in favor of a more geometrical treatment, reminiscent of Plato's. [...] While Ptolemy agreed with Plato's point of view, that the common terminology is grossly misleading, he offered a much simpler, and completely different, explanation of common usage than did Plato." Anna De Pace suggested that Ptolemy echoed *Tim.* 63a. She was not aware of Taub's work (see De Pace 2009, p. 122 and n. 369). See also Omodeo & Tupikova 2016, p. 168: "Additionally, [Ptolemy] affirms that there is no up-and-down motion in the universe, since directions depend on the observer. This statement is at odds with Aristotelian cosmology."

[69] "Nanque si diceremus, Grave est, quod deorsum manet, et leve quod sursum manet; non bene definiremus, cum sursum et deorsum non re, sed ratione tantum, distinguantur" (EN, I, p. 413).

[70] Taub 1993, p. 90, emphasis added. Peter Dear even claimed that "Ptolemy had commenced the *Almagest* with a brief treatment of the physical framework that constrained his accounts of the motions seen in the heavens, and it was a framework derived from the natural philosophy of Aristotle," and that "Ptolemy had used basically Aristotelian physical arguments in support of the doctrine of a central, stationary, spherical earth around which the heavens revolved" (Dear 2001, pp. 19, 41).

herself recognized when discussing *Alm.* I.5.[71] Physical arguments are introduced only to counter the idea of the Earth's daily rotation by the end of *Alm.* I.7. When it comes to its position in the universe, the arguments presented in the first book of the *Almagest* are rooted in observed astronomical phenomena. According to Ptolemy, these observations would not align with reality if the Earth occupied a noticeably off-center position. And if, as Taub suggested, Ptolemy built upon some of Aristotle's arguments from *De caelo*, it is clear that he did so with a meticulous approach, carefully selecting, refining, and structuring them.[72]

Notably, Ptolemy avoids falling into the circular argument also presented in a passage from the *Juvenilia*:

> [an argument for the Earth's centrality is] from Regiomontanus in the *Epitoma*, first book, [the end of the] third conclusion, and from Aristotle in *De caelo*, second book: all heavy things, when freely descending along the diameter of the world, meet the surface of the Earth at equal angles, no matter which part of the sphere they descend upon; therefore, as a consequence, they tend towards the center of the Earth. Otherwise, they would not strike the Earth's surface at equal angles. As a result, since the diameters of the world, along which the heavy objects move, pass through

---

[71] See Taub 1993, pp. 71, 74, 78–79. On *Alm.* I.7, see Omodeo & Tupikova 2016, p. 167: "[Ptolemy] does not use the physical arguments against the motion of the Earth in *Almagest* I, 7 to rule out the possibility of a tiny central displacement of the Earth. Unlike Aristotle, he seems to regard these arguments as irrelevant for demonstration of the centrality of the Earth. Ptolemy argues that the fall of bodies can be regarded as a corollary of geocentrism instead of an argument for it." This had already been emphasized by Pedersen: "The proof [for the Earth's centrality in the world] is indirect since Ptolemy shows that any other position is impossible for astronomical reasons, *leaving out any 'physical' argument derived from the theory of gravitation*" (Pedersen 2011, p. 39, emphasis added).

[72] See Taub's conclusion in Taub 1993, pp. 99–100. Pedersen, too, noted that "Aristotle begins with a series of physical arguments concerning gravitation, having previously proved the Earth to be at the center of the universe. This latter opinion is first established by Ptolemy in the following chapter [viz. *Alm.* I.5]. The proofs based on theories of gravitation have to wait, and he is left with purely astronomical arguments. This gives his treatment more unity and coherence than the corresponding chapter in Aristotle" (Pedersen 2011, pp. 38–39). It must be noted that Ptolemy's line of reasoning in *Alm.* I.5 bears a closer resemblance to that of Cleomedes in his *Meteora*, I.6, than to that of Aristotle in *De caelo*, II.14. According to Robert Balfour, Ptolemy took ("usurpa[vi]t") some of the arguments in the first book of the *Almagest* from Cleomedes (see Cleomedes 1607, pp. 47–49, and Belfour's commentary on pp. 209–215). For further details, see the notes in Bowen & Todd 2004, pp. 74–77.

the center of the universe, intersecting there, the center of both the Earth and the world is the same.[73]

In the *Juvenilia*, Ptolemy's name frequently appears alongside Regiomontanus. However, in this particular instance, the argument is ascribed only to Regiomontanus and Aristotle. The primary goal of this argument is the previously mentioned effort to deduce the Earth's centrality based on the movement of heavy objects. The circular reasoning, or *petitio principii*, hinges on the assumption that heavy objects naturally fall along the diameters of the celestial sphere. This fallacy might have been overlooked by many, perhaps misconstruing it as empirical evidence. Peurbach and Regiomontanus refer to this as a "direct argumentation" and go on to state that "*we see* heavy objects freely descending along the world's half-diameter."[74] They seem to be aware that this argument is foreign to Ptolemy. In fact, they opt to append it at the end ("*Possumus preterea idem directa argumentatione affirmare…*") after recapitulating the contents of *Alm.* I.5. This awareness is clear in the *Juvenilia*, where the argument is not attributed to Ptolemy.

In June 1971, Alistair C. Crombie discovered that the original source (though not necessarily the direct one) of some of the information found in the *Juvenilia* is the 1581 edition of the commentary on the *Sphere* by Clavius.[75] Indeed, Clavius does not attribute the argument to Ptolemy either. He further suggests that Regiomontanus actually derived the argument from Aristotle:

---

[73] "[…] ex Regiomontano in Epitome lib<ro> p<rim>o conc<lusione> 3.a, et ex Atistotele in 2.o Coeli, omnia gravia, libere secundum mundi diametrum descendentia, superficiei Terrae ad angulos aequales occurrunt, in quacunque orbis parte descendant; ergo, ut consequens est, tendunt ad Terrae centrum, alias non inciderent superficiei Terrae ad angulos aequales; quo fit ut, quia diametri mundi, secundum quas gravia feruntur, transeunt per centrum universi ibidem se intersecantes, ut idem sit et Terrae et mundi centrum" (EN, I, p. 48). For another English translation of this passage, see Wallace 1977, p. 72. For a summary of the six arguments exposed in the *Juvenilia*, see Wallace 1984a, p. 34.

[74] "*Videmus* enim gravia libere secundum mundi semidiametrum descendentia" (Peurbach & Regiomontanus 1496, emphasis added).

[75] See Carugo & Crombie 1983, p. 6.

The third reason is from Johannes Regiomontanus in the *Epitoma*, book I, conclusion 3, *which he seems to have taken from Aristotle*'s *De caelo*, book II.[76]

It is now crucial to highlight that the initial section of *Alm.* I.7 is open to an interpretation distinct from those posited by Schreckenfuchs, Reinhold, and even earlier by Theon, or more recently by Taub. That is, Ptolemy's rationale for geocentrism—and, as consequently, his view on gravity—was not necessarily always linked to Aristotle's *De caelo*. Indeed, for Galileo, it appears this was not the case.

GALILEO'S NON-ARISTOTELIAN PTOLEMAIC COSMOLOGY

A preliminary note in *De motu antiquiora* reads:

> *It is possible to ask whether heavy objects truly move towards the center*; on this matter, Ptolemy discusses in chapter 7 of the first book of the *Almagest*.[77]

Given that Galileo is still referring to *Alm.* I.7 here, as in the treatise, this might be a notation taken earlier to remind himself of the precise passage from the *Almagest* to cite later in the treatise. However, I believe that, even though the reference to the *Almagest* is identical, in this context, Galileo emphasizes a slightly different matter.[78]

In this note, Galileo questions whether heavy bodies truly move toward the center of the Earth and, thus, of the world, rather than accepting this notion outright. He believes it is valid to challenge well-established views on motion. This skepticism is tied to the *Almagest*. From Ptolemy's masterpiece, Galileo does not merely adopt the idea that heavy objects naturally move toward the center of the world. He also learns to question this idea, realizing that Ptolemaic cosmology does not derive its validity from Aristotle's doctrine of motion, but rather it provides evidence that

---

[76] "Tertia ratio est Ioannis Regiomontani in Epitome lib. 1. concl. 3. *quam sumpsisse videtur ex Aristotele* lib. 2. de caelo" (Clavius 1581, p. 143, emphasis added).

[77] "*Quaeri potest, an gravia vere ad centrum moveantur*; de quo Ptolemaeus, c<apitulo> 7. p<rim>i lib<ri> Al<magesti>" (EN, I, p. 418, emphasis added).

[78] This was likely Drabkin's interpretation of the passage, as he emphasized that the question raised in this note was not addressed in the treatise. See Galilei 1960, p. 130.

heavy objects are drawn to the Earth's center and, by extension, the center of the universe.

That Galileo interprets *Alm.* I.7 in this manner, differentiating Ptolemy's view from Aristotle's, is evident from a previously mentioned passage in the *De motu antiquiora* treatise:

> Aristotle seems to suggest something similar in *Physics* VIII, text 32, when, *asking why heavy and light objects move to their proper places*, he supposes that the reason is because they are naturally inclined to move somewhere – the light objects upward and the heavy ones downward. *However*, Ptolemy, at the beginning of the seventh chapter of the first book of his *Great Construction*, says *it is vain to inquire why* heavy objects are drawn to the center [of the universe], having demonstrated that *the Earth, to which they are drawn*, is in the center. But such demonstrations [*haec*] do not remove the difficulty: *granting that heavy objects are drawn to the center* [of the universe] *because they are drawn to the Earth*, we further ask why the Earth was placed in the center and not in the place of fire.[79]

First of all, it must be noted how Galileo emphasizes the contrast between Aristotle's and Ptolemy's views. He points out that Aristotle raised the question ("… asking why…") about the motion of heavy and light objects, which Ptolemy considered vain and useless ("… it is vain to inquire why…"). Moreover, in my translation, the plural neutral demonstrative pronoun "*haec*" is understood in relation to Ptolemy's demonstrations: (*a*) those asserting that the Earth is at the center of the universe (*Alm.* I.5–6); (*b*) the one stating that heavy objects move toward the Earth's center (*Alm.* I.7). However, since Galileo references Aristotle and other "philosophers" before mentioning Ptolemy, the "*haec*"

---

[79] "Simile quiddam Aristoteles, 8 Phys<icorum> 32, adferre videtur, dum, *quaerens cur gravia et levia ad propria moveantur loca*, subdit, causam esse quia habent a natura ut sint apta ferri aliquo, et hoc leve quidem sursum, grave autem deorsum. Ptolemaeus *autem*, in principio 7i cap<ituli> p<rim>i libri suae Magnae Constructionis, inquit *frustra inquiri cur gravia ad medium ferantur*; cum demonstrasset *Terram, ad quam feruntur*, in medio esse. Verum haec difficultatem non tollunt: *dato, enim, ferri ad medium quia ad Terram ferantur*, rursus cur Terra in medio non autem in loco ignis posita fuit, quaerimus" (EN, I, pp. 344–345, emphasis added).

has been interpreted as also pointing to the reasons of all these philosophers.[80] I believe this interpretation is unfounded, because Aristotle would never have conceded that heavy objects move toward the center of the world as a consequence of the fact that they move toward the center of the Earth, as Galileo explicitly states shortly after. It is clear that Galileo is critically engaging with Ptolemy's arguments.[81] If heavy objects move toward the center of the Earth, and by extension, toward the center of the world due to the Earth's central position, then the question arises: why is the Earth indeed at the center of the world?

So, upon reading the *Almagest*, Galileo felt it was reasonable to question if heavy bodies truly moved toward the center of the world. For Ptolemy, after proving the Earth's central position with astronomical evidence and showing that heavy bodies fall toward its center, there was no need to further question why *they* were drawn to the center of the world. Galileo accepted Ptolemy's demonstrations, but then posed a cosmological question Ptolemy had bypassed: why is the *Earth* at the center of the universe in the first place? To an Aristotelian philosopher, this might seem nonsensical. However, it becomes a pertinent inquiry for those who embrace Ptolemy's arguments for the Earth's centrality while dismissing both Aristotle's physical and cosmological explanations from *De caelo*.

Indeed, in *De motu antiquiora*, Galileo showcases his adherence to a Ptolemaic and yet non-Aristotelian geocentric view. He posits that the Earth's central position in a spherical universe is not arbitrary.[82] The

---

[80] See the passage quoted *supra* in this chapter, n. 24. Fredette offered a translation, suggesting between square brackets that "*haec*" pertains to all the previously explained arguments: "But these [arguments] do not remove the difficulty."

[81] This is further evidenced by the fact that Galileo included the phrase "*verum haec difficultatem non tollunt…*" in the second version of the treatise, where he also references Ptolemy's *Almagest*. In contrast, the first version, which only reports the opinions of philosophers and Aristotle, does not contain this passage.

[82] This is evident in all versions of the treatise. In the following quotation words from the three different versions are indicated with a superscripted number (e.g., "Quod$^{2-3}$" indicates that this Latin word appears both in the second and third version). Words without a superscripted number are common to all versions. A slash between two words indicates substitution. The word before the slash was replaced by the one after. If one or more words are included between square brackets, this serves to indicate where they are placed in the respective version(s). Additionally, a line between different punctuation signs indicates changes in punctuation. Henceforth, I shall use this transcription system whenever I deem it appropriate. "Attamen,^{1}/Quod$^{2-3}$ si rem accuratius spectemus, non erit

rationale derives from the property of a sphere, wherein concentric spaces closer to the center are "narrower" (have less volume) than the "broader" spaces farther out. For Galileo, the order of the universe corresponds in a congruent manner to the inherent nature of the four elements within it. He contends that all matter is made of the same substance, with denser objects having more of this substance per volume. For instance, among the elements earth, water, air, and fire, the densest—i.e., earth—would take up the least volume, followed by water, then air, and finally fire. This density-driven hierarchy matches the world's arrangement, with denser elements like earth being closest to the center of the universe.[83]

In *De motu antiquiora*, Galileo postulates a natural origin for elements based on mathematical relationships between matter and volume. This idea proposes a universe where elemental placement aligns with inherent elemental characteristics. So, while Galileo's cosmic order might resemble Aristotle's, his understanding of the "form" (*forma*) of the elements is distinct: gone is Aristotle's concept of absolute heaviness. Instead, Galileo

---

profecto existimandum, nullam in tali distributione necessitatem aut [saltem]³ utilitatem habuisse naturam, sed solum ad libitum et casu quodammodo operatam esse¹/fuisse²⁻³. Hoc cum de provida natura nullo [posse]³ pacto existimari [posse]¹⁻² perpenderem, [in excogitanda]²⁻³ [interdum anxius fui]¹ [in excogitanda]¹, nisi necessaria, saltem [utili et]²⁻³ congruente [ac utili]¹, aliqua causa [interdum anxius fui]²⁻³" (EN, I, pp. 252, 342–343, 345). In the dialogued version, Alexander provides a *ratio* for the order of nature. Dominicus claims that the *ratio*, although not *potissima*, is to be deemed verisimilar: "Ratio ista, quamvis talis elementorum dispositionis existimanda non sit potissima, attamen nonnullam in se habet veritatis speciem, cui animus libenter assentitur" (EN, I, p. 364).

[83] See, for instance, the *ratio* posed by Alexander in the dialogued version and mentioned in the previous note: "Nisi forte velimus dicere, graviora centro propinquiora esse quam leviora, quia videntur quodammodo ea esse graviora, quae in angustiori loco plus materiae continent: ut, verbigratia, si fuerit saccus lana plenus, quae in eo nulla vi sit constipata, deinde magna cum violentia multo plus lanae in eodem comprimatur, tunc gravior erit quam antea, quia in eodem spatio plus materiae cumulabitur. Cum, itaque, spatia quae centro mundi sunt propinquiora, sempre angustiora sint iis quae a centro magis recedunt, rationi consentaneum fuit ut ea replerentur materia, cuius maior gravitas, quam alterius, angustiora spatia occuparet." The illustration of the sack full of wool is omitted in all three treatise versions, but the order of nature is attributed to the same *ratio*: "[in sphaera]²⁻³ angustiora sunt loca [in sphera]¹ quo magis ad centrum accedimus¹/accedunt²⁻³, ampliora vero quo [magis]¹ ab eodem [magis]²⁻³ recedimus¹/distant²⁻³: prudenter, igitur, simul et aeque terrae statuit natura locum esse qui caeteris est angustior, nempe prope centrum; reliquis deinde elementis loca eo ampliora, quo ipsorum materia rarior esset" (EN, I, pp. 253, 343, 345).

leans toward Archimedean views on weight. This means that qualitative differences between the four elements are reduced to the ratios of a certain quantity of matter and the form (i.e., volume) of each element. In other words, differences in the kinds of bodies are reduced to differences in density. Thus, the characteristics that distinguish one type of body and define its place in the world can be understood in terms of how densely the matter is packed or distributed.[84]

In essence, Galileo's *De motu antiquiora* attempt to reconcile Archimedean physics with Ptolemaic geocentrism, suggesting a geocentric worldview that diverges from the Aristotelian one. He envisions an Archimedean-Ptolemaic cosmology where a cosmogony narrative bridges both schools of thought.[85]

---

[84] The premise is the same in all treatise versions, although in the first one it is attributed to the Atomists. That is, there is only one kind of matter, and heavier bodies are those composed of more matter per volume. From this, it follows that "rationi profecto consentaneum fuit, ut quae in angustiori loco$^{1-2}$/spatio^3 plus materiae concluderent, angustiora etiam loca, qualia sunt quae centro magis accedunt, occuparent. Ut si, exempli gratia, intelligamus, naturam in prima mundi compagine totam elementorum communem materiam in quatuor [aequas]^1 partes divisisse,^{1}|;$^{2-3}$ deinde ipsius terrae formae suam materiam tribuisse, itidem et formae aëris suam,^{1}|;$^{2-3}$ terrae autem formam materiam suam in angustissimo loco constipasse, aëris autem formam in amplissimo loco materiam suam reposuisse,^{1}|;$^{2-3}$ nonne congruum erat ut [aëri]$^{2-3}$ natura [aëri]^1 magnum spatium [assignaret]$^{1-2}$, terrae autem minus [assignaret]^3?" (EN, I, pp. 253, 343, 345). As noted by Maarten Van Dyck, in this passage, the term *forma* has "taken on an entirely geometrical character" (Van Dyck 2006, p. 152 n. 441).

[85] I refer here to the cosmogony inserted in the third version of the treatise: "Vastissimae caelestis excrementa sphaerae, post illius mirabilem compaginem, divinus Opifex, ne forte immortalium beatorumque spirituum offenderent intuitum, in eiusdem globi centrum extrusit atque abscondidit: verum, cum satis amplum et capax sub ultimi concava superficie orbis relictum spacium densissima gravissimaque illa materia mole sua non expleret, ne magnum spacium otiosum atque vacuum esset, quae, pressa gravitate sui, onerosam illam indigestamque massam, in angustis se cancellis concluserat, distraxit; et ex illius innumeris particulis plus minusve rarefactis quatuor illa efformavit corpora, quae postea elementa diximus. Quorum quod gravissimum densissimumque, ut prius erat, remansit, e loco in quem antea confugerat non removit; et sic relicta est terra in centro: et, simili ratione, quae densiora fuerant, terrae viciniora constituta sunt. Eorum vero quae ex hac materia constituta sunt corpora, densiora illa dicta sunt quae, sub eadem mole, plures eiusdem materiae particulas coëgere; densiora, autem, graviora fuere" (EN, I, p. 344). See Galluzzi 2011, pp. 14–15, 17 n. 18. Galileo's narrative here is based on commonplaces finely rearranged (see Malara 2021, pp. 214–219).

Conclusion

In *De motu antiquiora*, Galileo seems to accept Ptolemy's astronomical proofs for geocentrism. At the same time, he rejects Aristotle's doctrine of natural motion and gravity. This peculiar combination of beliefs leads him to seek a new geocentric cosmology. In doing so, Galileo is aware that he is searching for a cosmological explanation that Ptolemy explicitly disregarded.

A distinctive aspect of Galileo's thought in *De motu antiquiora* is found in his attempt to create a synthesis that challenges Aristotelian concepts by incorporating different traditions, such as the Ptolemaic and Archimedean. By addressing the natural order of the elements in the world and applying hydrostatic principles to the study of motion, Galileo also aimed to advance beyond the teachings of these ancient mathematicians.

Although the separation of Ptolemaic arguments from Aristotelian ones in favor of geocentrism might seem innovative, it aligns with the practices of Galileo's time. This awareness is also evident in one of his later work, the *Dialogue*. In the Second Day, Simplicio distinguishes between the arguments of Aristotle and those of Ptolemy concerning the Earth's centrality and immobility.[86] Salviati concurs with Simplicio's distinction and adds that

> The arguments presented on this matter fall into two categories: some relate to earthly phenomena, without any connection to the stars, while others are drawn from the appearances and observations of celestial things. Aristotle's arguments are mostly derived from things around us, leaving the others to the astronomers; therefore, it would be good, if you agree, to first examine those derived from experiences on Earth, and then we will move on to the other category.[87]

---

[86] See EN, VII, pp. 150–151.

[87] "Gli argomenti che si producono in questa materia, son di due generi: altri hanno riguardo a gli accidenti terrestri, senza relazione alcuna alle stelle, ed altri si cavano dalle apparenze ed osservazioni delle cose celesti. Gli argomenti d'Aristotele son per lo più cavati dalle cose intorno a noi, e lascia gli altri alli astronomi; per[ci]ò sarà bene, se così vi pare, esaminar questi presi dalle esperienze di Terra, e poi arriveremo all'altro genere" (EN, VII, p. 151).

However, such a distinction did not imply incompatibility. If scholars like Clavius, or even earlier figures such as Alessandro Piccolomini and Philip Melanchthon,[88] already had a clear understanding of the difference between the arguments presented by Aristotle and those put forth by Ptolemy in support of geocentrism, none of them denied their complementarity. Some, such as Schreckenfuchs and Reinhold, along with Theon in antiquity, went so far as to integrate both Aristotelian and Ptolemaic arguments in favor of the Earth's centrality. There were also those who, assuming that Ptolemy adhered to Aristotle's views, were surprised to discover that both he and Theon deviated from Aristotle's teachings. I refer here to Francesco Buonamici, who interpreted a passage from *Alm.* I.7 as presenting a theory similar to that of Anaximander. According to Anaximander, as interpreted by Aristotle, the Earth is stable at the center because all its parts are equally distant from the periphery of the world. However, Buonamici, following Aristotle, held that this theory was fallacious[89]:

> For this reason, we marvel that here Ptolemy, an admirer of Aristotle, and Theon, his disciple, assigned a similar [to Anaxagoras'] cause of stability [to the Earth].[90]

Highlighting the contrast between Aristotle's views and those of Ptolemy was not news. This contrast had already been brought to light by Arab philosophers and astronomers.[91] It persisted in various Renaissance

---

[88] As noted by Emil Wohlwill, in the 1549 and 1550 edition of the *Initia doctrina physicae*, the arguments of Ptolemy and Aristotle are presented in the chapter on "What is the motion of the world?" (see Wohlwill 1904, p. 262). See also the English translation in Omodeo & Regier 2019, pp. 105–108, esp. 106–107 for the arguments laid out in *Alm.* I.5, and p. 108 for the argument in *Alm.* I.6.

[89] See *De caelo*, II, 295b 11–25. Buonamici clearly expands on Aristotle's critique. He says that Anaximander's argument "peccat etiam fallacia consueta secundum non caussam ut caussam" (Buonamici 1591, p. 455A). On Ptolemy's argument in favor of the Earth's immobility, see Pedersen 2011, pp. 43–44, notwithstanding anachronistic terminology (see Omodeo & Tupikova 2016, p. 169, n. 68).

[90] "Quamobrem est, quod hic Ptolemaeum Aristotelis admiratorem et Theonem illius alumnum consimilem caussam quietis assignasse miremur" (Buonamici 1591, p. 455B).

[91] Averroes' philosophical critiques of Ptolemy are widely renowned: see Forcada 2014. For a useful and well-written overview on this matter, see Lerner 1996, vol. I, pp. 99–110.

attempts to eliminate epicycles, deferents, and equants, along with some efforts to make the homocentric system a valid alternative.[92]

However, what I have not managed to find in other authors so far is an interpretation of Ptolemy's geocentrism as a reversal of the Aristotelian one. In the *Almagest*, Galileo encountered a theory suggesting that heavy bodies move toward the center of the Earth and, consequently, toward the center of the world. For if the Earth were not at the center of the world, we could not observe the phenomena in the sky that we witness daily. The reversal lies in the fact that, according to Aristotle, heavy bodies accidentally move toward the Earth but by necessity toward the center of the world. In contrast, according to Galileo's reading of Ptolemy, heavy bodies move accidentally toward the center of the world but by necessity toward the center of the Earth. Only within this latter perspective does it make sense to contemplate why the Earth is at the center, and not, for instance, in the place occupied by the element of fire. While Ptolemy shows indifference to this problem, Galileo insists on the conviction that nature does not operate randomly. At the very least, humans can attain a verisimilar understanding of the underlying *ratio* that governs the order of nature.

In conclusion, in *De motu antiquiora* Galileo engages closely with the first book of the *Almagest*, avoiding the conflation of the Aristotelian doctrine of motion with Ptolemaic reasoning. One might argue that Galileo, influenced by Ptolemy and critical of Aristotle, aimed to provide a solid physical foundation for the astronomy presented in the *Almagest*. This inclination appears to anticipate his later involvement with Copernicus' *De revolutionibus*. However, caution is warranted as this interpretation likely involves reading too much into the few passages of *De motu antiquiora* that have been considered thus far.

---

[92] On Regiomontanus' interest in homocentric astronomy, see Shank 1998, Swerdlow 1999, and Shank 2002, pp. 185–188. On the homocentric views developed by Giovan Battista Amico and Girolamo Fracastoro, see Di Bono 1990, Di Bono 2006. More technical insights in Swerdlow 1972, and Di Bono 1995. Numerous works provide a general overview of the status of homocentric astronomy after the reception of the *Almagest*. I will limit myself to mentioning three, which also include references to other literature on the subject: see Kren 1968; Lerner 1996, pp. 111–115; Omodeo 2014, pp. 77–79.

CHAPTER 4

Galileo as a Commentator on *Almagest* I.3

Abstract The fourth chapter focuses on Galileo's critique of the traditional interpretation of Ptolemy's theory on why stars appear larger at the horizon, as hinted in *Alm.* I.3. Galileo's theory, also seen in *The Assayer* (1623) and Alimberto Mauri's *Considerations* (1606), is similar but not identical to the one found in Theon's commentary. Although it is unclear if Theon was a direct source, the chapter shows Galileo critically engaging with the *Almagest* and challenging conventional views.

Keywords Galileo Galilei · Claudius Ptolemy · Theon of Alexandria · *Almagest* · *De motu antiquiora* · Moon illusion

On one occasion, it appears possible to infer from *De motu antiquiora* how Galileo interpreted a specific passage in Ptolemy's masterpiece, namely *Alm.* I.3.[1]

---

[1] Libero Sosio was the first to pinpoint this specific reference to the *Almagest* (see Galilei 1965, pp. 277–278 n. 513). In the sixteenth century, there were various editions of the *Almagest* circulating. In some versions, the chapter in discussion is listed as the second, not the third. However, for clarity and ease, I am using the chapter division laid out by Toomer, as already explained *supra*, in the beginning of Chapter 3.

© The Author(s), under exclusive license to Springer Nature Switzerland AG 2024
I. Malara, *Galileo and the* Almagest, *c.1589–1592*, Palgrave Studies in the History of Science and Technology,
https://doi.org/10.1007/978-3-031-70614-1_4

This passage addresses the debate on the shape of the celestial vault. Some argued against its sphericity, perhaps also by noting that celestial bodies appear larger near the horizon and smaller at the zenith. This might indicate a variable distance from the observer to the stars, thus challenging the idea of a spherical celestial vault.[2]

In the *Almagest*, Ptolemy proposed that the change in apparent size is due to optical effects, specifically those caused by the presence of vapors and exhalations between the observer and the celestial body. He likened this to observing an object submerged in water: from an external viewpoint, the deeper the object, the larger it appears. Notably, Gerard of Cremona's Latin translation contains this analogy but omits the mention of vapors[3]:

---

[2] In the third chapter of the first book of the *Almagest*, Ptolemy argues for the spherical shape of the heavens from the apparent revolution of the stars. See Pedersen 2011, pp. 36–37, and Toomer 1984, pp. 38–40.

[3] The whole passage in Toomer's translation from the Greek runs as follows: "To sum up, if one assumes any motion whatever, except spherical, for the heavenly bodies, it necessarily follows that their [viz. of the stars] distances, measured from the earth upwards, must vary, wherever and however one supposes the earth itself to be situated. Hence the sizes and mutual distances of the stars must appear to vary for the same observers during the course of each revolution, since at one time they must be at a greater distance, at another at a lesser. Yet we see that no such variation occurs. For the apparent increase in their sizes at the horizons is caused, not by a decrease in their distances, but by the exhalations of moisture surrounding the earth being interposed between the place from which we observe and the heavenly bodies, just as objects placed in water appear bigger than they are, and the lower they sink, the bigger they appear" (Toomer 1984, p. 39).

Tr. Gerard of Cremona (Venice 1515)	Tr. George of Trebizond (Venice 1528)	Tr. Erasmus Reinhold (Basel 1549)
Although we see [the stars] with an increase in their magnitude when they are at the horizon, this is not shown to us by their proximity and shorter distance [*parvitas longitudinis*] from the horizon. It is almost similar to what is placed in water: it appears larger [than it really is], and the more it is immersed in depth, the more its [apparent] magnitude increases[4]	The fact that stars appear larger near the horizon is not due to a shorter distance, but rather to the *exhalation rising from the earth in such a way that it interposes itself between our sight and the stars themselves*, much like submerged objects that seem larger the deeper they go[5]	The fact that stars appear larger near the horizon is not due to a shorter distance, but rather *because of vapors in the air that are between our sight and the stars*, just as submerged objects appear larger in water, and the deeper they are immersed, the larger they seem[6]

The omission of the 1515 edition was not significant during Galileo's time since Ptolemy's vapor explanation was widely cited in other works.

The Standard Interpretation

In his *De Sphaera*, Sacrobosco—mainly inspired by al-Farghani—credits Ptolemy with the idea that vapors, denser than air, are responsible for the optical magnification seen when celestial bodies are near the horizon.[7] By the sixteenth century, this interpretation was commonly associated with

[4] "Quod autem videmus ex augmento magnitudinis earum cum sunt apud horizontas non demonstrant nobis sic esse propinquitas earum et parvitas longitudinis earum ab horizonte. Sed est quasi simile ei quod in aqua ponitur. Videtur enim maius, et quanto plus in profundum mergitur, augmentatur eius magnitudo" (Ptolemy 1515, f. 3r).

[5] "Nam quod iuxta horizontes maior magnitudo stellarum videatur, non distantiae parvitas id facit, sed *huiusmodi terra obeuntis evaporatio quum inter visum nostrum et stellas ipsas exhalet*; veluti maiora in aquis submersa videntur, et quidem tanto maiora quanto profundiora petierint" (Ptolemy 1528, f. 2r, emphasis added).

[6] "Nam quod iuxta Horizontes stellae videntur maiores, id non fit propter distantiam breviorem, sed *propter vapores in aëre, qui inter nostrum visum et stellas existunt*, sicut et maiores videntur res in aqua mersae, et quo profundius merguntur, eo maiores apparent" (Ptolemy 1549, f. 49v, emphasis added).

[7] See Thorndike 1949, pp. 81 (Latin), 120–121 (English). For the reference to al-Farghani, see al-Farghani & al-Battani 1537, pp. 3r–v. For another work, undoubtedly familiar to Galileo, in which the same theory is presented, see Barozzi 1585, p. 17. In

Aristotle's descriptions in *Meteorologica* and the optical theories of figures like Alhacen and Witelo.[8] Other classical scholars, including Cleomedes, Alexander of Aphrodisias, Seneca, and Macrobius, were sometimes cited in this context.[9] Citing all these authors (and authorities) served to convey the belief that the magnification effect resulted from refraction caused by a medium denser than air, positioned between the observer and the celestial body. Today we know that this explanation is inaccurate; it is an illusory phenomenon. Notably, even back then, a few were

this context, the theory is employed as the "fourth argument" (*quarta ratio*) in support of the spherical shape of the world. In the margin, Barozzi notes that this is a beautiful argument ("ratio... pulcherrima") of Ptolemy, which is not treated conclusively by al-Farghani and Sacrobosco ("... in qua Alfraganus et Io. de Sacr. nugantur"; translated as follows in Barozzi 1607, p. 67: "... nella quale Alfragano e Gio. di Sacrobosco non conchiudono"). On the contrary, Theon provides its geometric demonstration (see *infra*, n. 24 in this chapter).

[8] See *Meteor*. III.4, 373 b 11–14 (see also *Problemata*, XXVI, 53, 946 a 33–34). As for Alhacen and Witelo, the reference was usually to Alhacen's *De aspectibus*, end of book VII, (see Smith 2010, vol. I, p. 143 (Latin); vol. II, pp. 330–331 (English translation), par. [7.73]), and to Witelo's *Perspcetiva*, book X, prop. 54: "Omnes stellae videntur rotundae, maiores [*sic*; it should be *minores*, as already noted by Kepler] in horizonte quam in medio coeli, nisi quandoque contrarium accidat propter interpositos vapores visibus et stellis." These two works were bound together in the famous *Opticae thesaurus*, edited by Friedrich Risner and published in 1572 (see Alhacen & Witelo 1572, p. 282 of *De aspectibus*, and pp. 448–449 of *Perspectiva*). Galileo owned this book (see Favaro 1886, p. 262).

[9] See Cleomedes, *Meteora*, II.1, 27–44; Alexander of Aphrodisias, *Problemata*, I.36; Seneca, *Quaest. Nat.*, I, 3.9; 6.5; Macrobius, *Saturnalia*, VII.14. Except Cleomedes, all the authors mentioned in this and the previous note were cited, for instance, by Della Porta in his *De refractione* (1593), book I, prop. 11. See Smith 2017, pp. 36–41 (Latin and English translation).

aware of this.[10] Regardless, the two types of explanations were not always perceived as mutually exclusive.[11]

At its core, the dominant belief, not always explicitly linked to Ptolemy, was that thick layers of vapors or exhalations, which are denser than air, positioned themselves between the observer and the celestial body. This is how Clavius explained the whole argument in his commentary on the *Sphere*:

> [John of Sacrobosco] had said, following al-Farghani, that the Sun, Moon, or any other star appear larger near the horizon rather than overhead [i.e., at the zenith]. One could infer from this that the sky is not round, since it is not uniformly distant from all directions. For where a star appears larger, there the sky will be closer; but where it appears smaller, there it will be more distant. Therefore, John of Sacrobosco addresses a silent objection. He says that the reason the Sun, Moon, or any other star appear larger in the east and west [i.e., near the horizon] rather than in the middle of the sky, or zenith, is not that they are more distant from us there [at the zenith] than here [at the horzon], at least perceptibly so; but the reason is that the vapors lifted from the earth intervene between the Sun or any

---

[10] This is the case of Gemma Frisius, who showed that his *radius astronomicus et geometricus* do not detect any enlargement in the apparent diameter of celestial bodies near the horizon. See Frisius 1545, ff. 29r–v. According to Goldstein, this passage by Gemma is a critique to *Alm.* I.3, namely the passage discussed in the present chapter (see Goldstein 1987, pp. 173 (English translation), 170 (commentary); see also Dupré 2002, pp. 35–36). However, Gemma Frisius explicitly opposes the theory of those who adhered to the homocentric system and thus denied that the variations in the Moon's apparent diameter equate to changes in distance. They claimed that the variations in apparent diameter were merely apparent, determined by atmospheric refraction. Although Gemma Frisius does not name his opponents, it is highly likely that his targets were Giovan Battista Amico and Girolamo Fracastoro. See Amico 1536, pp. 15v–17r (on variations in the Sun's apparent diameter), 21r–22r (on variations in the Moon's apparent diameter), and Di Bono 1990, pp. 88–89, 95–97, 172–175, 187–189; Fracastoro 1538, pp. 18r–19r, and Reeves 2008, pp. 32–34. See also Caverni 1891–1900, vol. I, p. 45. A brief mention also in Omodeo 2014, p. 83.

[11] They are found and coexist in Cleomedes and Alhacen. See Cleomedes, *Meteora*, II.1, 27–75; Ross 2000; Bowen & Todd 2004, pp. 101–103. On Alhacen, see Alhacen & Witelo 1572, *De aspectibus*, p. 282; Smith 2010, vol. I, pp. 141–144 (Latin), vol. II, pp. 328–331 (English), pars. [7.70–74]. For Alhacen, the optical magnification is a secondary cause, which depends on the presence of a band of vapors denser than air between the observer and the star. On the other hand, he believed that "the primary and invariant (*permanens*) cause lies in how we perceive distance and size" (Smith 2003, p. 110). The reader would benefit greatly, both in terms of understanding and reference to the literature, from reading Smith's notes 193–195 in Smith 2010, vol. II, pp. 394–397.

other celestial body and our sight. Hence, these vapors, being denser and thicker near the horizon, alter our visual rays, and as a result, we do not perceive the object in its true size. This is clearly evident, as [John of Sacrobosco] says, in a coin placed at the bottom of clear and transparent water.[12]

This theory gains further support through the authoritative reference to the *Perspectivi*, namely Alhacen and Witelo:

Al-Farghani proposes the same cause in the second book [of *Rudimenta astronomica*], and the *Perspectivi* demonstrate it. For from that variation of visual rays, any object appears closer, and therefore larger. For the same reason, it happens that an object sometimes appears to us through refracted rays, which otherwise could in no way reach our eyes through direct rays. We have a very clear example in a coin thrown at the bottom of a vessel of moderate depth. If we move backward until we cannot see that coin due to the intervening sides of the vessel, and if the vessel is then filled with clear water, that coin will immediately appear and present itself to our sight. Hence, it sometimes happens that the Sun, Moon, and other stars appear to us before they rise above the horizon.[13]

---

[12] "Dixerat [Ioannes de Sacrobosco] in ratione Alphragani, Solem et Lunam, aut quamcunque aliam stellam maiorem apparere iuxta Horizontem, quam supra verticem capitis; posset aliquis hinc inferre, caelum non esse rotundum, quandoquidem non aequaliter undique distat. Ubi enim stella maior apparet, ibi caelum propinquius existet; ubi vero minor, ibi remotius. Idcirco occurrit [Ioannes de Sacrobosco] tacitae huic obiectioni, dicens, causam cur Sol vel Luna, aut alia stella maior appareat in ortu et occasu, quam in medio caeli, seu vertice, non esse quod magis ibi, quam hic distet a nobis, saltem sensibiliter; sed esse vapores a terra elevatos, qui interponuntur inter Solem, vel quodlibet aliud astrum, et visum nostrum. Unde fit ut vapores illi, cum sint iuxta Horixontem spissiores, crassioresque, varient nostros radios visuales, et propterea minime cernamus rem in sua propria quantitate. Quod quidem evidenter patet, ut ait [Ioannes de Sacrobosco], in denario aliquo in fundo aquae perlucidae, atque clarae" (Clavius 1581, pp. 107–108).

[13] "Hanc eandem causam affert Alphraganus differ. 2 eamque demosntrant Perspectivi. Nam ex illa variatione radiorum visualium res quaevis propinquior apparet, unde et maior. Eadem de causa contingit rem aliquam videri per radios aliquando refractos, quae alias per directos ad oculum nostrum pervenire nequaquam potest. Exemplum clarissimum habemus in denario aliquo proiecto in fundo alicuius vasis vacui mediocris altitudinis. Si enim eo usque retrocedamus, donec denarium illum ob interiecta latera vasis inter ipsum et nostrum visum videre nequeamus, deinde vero vas illud repleatur [*sic*] aqua limpida, subito apparebit denarius ille, atque conspectui nostro sese offeret. Hinc denique fit, nonnunquam Solem, Lunam, et reliquas stellas apparere nobis, antequam supra Horizontem ascenderit" (ibid., p. 108).

So, Clavius applies the explanation based on vapors not only to account for the variations in apparent diameter but also to explain the displacement of celestial bodies. In both cases, the layers of vapor were supposed to be at a specific distance and height above the ground. This supposition explained why celestial bodies did not seem larger, or displaced, when directly overhead.[14]

Ptolemy's comparison to an object that appears larger the deeper it is submerged in water was not frequently detailed. Clavius' reference to the coin was commonplace. It was also utilized by Alhacen and Witelo.[15]

Occasionally, visual aids were used to explain this comparison. For example, Schreckenfuchs used a specific illustration in his notes on the first book of the *Almagest*. However, its effectiveness is open to interpretation (see Fig. 4.1). Barozzi, too, showcased an illustration in his *Comsographia* (see Fig. 4.2). Within this depiction, it becomes apparent that vapors construct a kind of wall separating the observer from the celestial body. Barozzi exploits the optical effect of vapors as a fourth supporting argument (*"ratio"*) for the sphericity of the celestial vault. Barozzi's figure effectively captures Alhacen's point that the vapor surface facing the observer must be to be planar.[16] However, both Schreckenfuchs and Barozzi used figures that were already part of the *Sphere* tradition.[17]

Theon's Interpretation

While not widely recognized or often referenced, there was another interpretation of Ptolemy's words. In his commentary on the *Almagest*, Theon clarifies Ptolemy's analogy of objects submerged in water by referencing a treatise on catoptrics, which he attributes to Archimedes.[18] In this treatise, Archimedes is said to have demonstrated that objects submerged

---

[14] See Fig. 8 in Lehn & van der Werf 2005, p. 5631.

[15] As Elio Nenci suggested to me, Clavius' sources were most likely Alhacen and Witelo. See, for instance, Alhacen & Witelo 1572, pp. 253 (*De aspectibus*, book VII, prop. 17), 414 (*Perspectiva*, book X, prop. 1).

[16] See Smith 2003, p. 110.

[17] See Taub 1993, p. 50, Fig. 3.4.

[18] Unfortunately, Theon's explanation is rarely mentioned in historical studies on the "Moon illusion." Cf. Ross & Ross 1976, p. 379, where the two scholars briefly mention Theon to highlight that his explanation is "obscure."

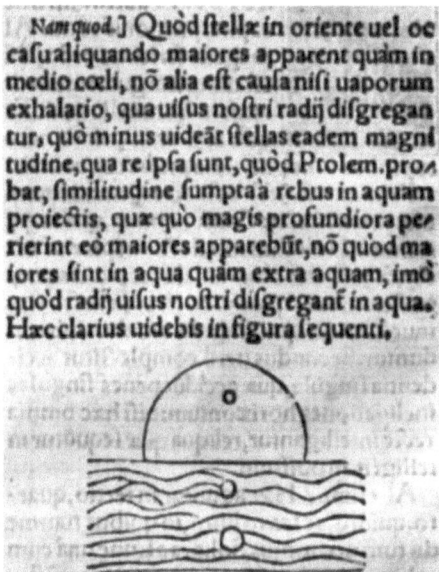

Fig. 4.1 Illustration from Schreckenfuchs' notes on the *Almagest* (see Ptolemy 1551). In English, the Latin passage appended to it reads as follows: "The reason why stars appear larger in the east and west [i.e., at the horizon] than they do in the middle of the sky [i.e. at the zenith] is none other than the exhalation of vapors. These vapors scatter the rays of our vision, preventing us from seeing the stars at the same size as they actually are. Ptolemy proves this, drawing a similarity from objects cast into water: the deeper they sink, the larger they appear. This does not mean they are larger in water than outside of it, but rather that the rays of our vision are scattered in water. This will be clearer in the following illustration"

in water appear larger, and the deeper they are, the larger they seem.[19] Thus, Theon brings to light the alleged geometric demonstration from Archimedes (see Fig. 4.3).[20]

[19] In Della Porta's translation: "Quod quemadmodum ea, quae iniiciuntur in aquam maiora videntur, et quanto infra subsident, eo maiora" (Ptolemy & Theon 1605, p. 18; Della Porta 2000, p. 31, ll. 288–290).

[20] Here, I will not delve deeper into the various interpretations of the "Moon illusion" during the Renaissance. For this reason, I have chosen not to emphasize specific details, otherwise important, such as Theon's adoption of the extramissionist view. For differences

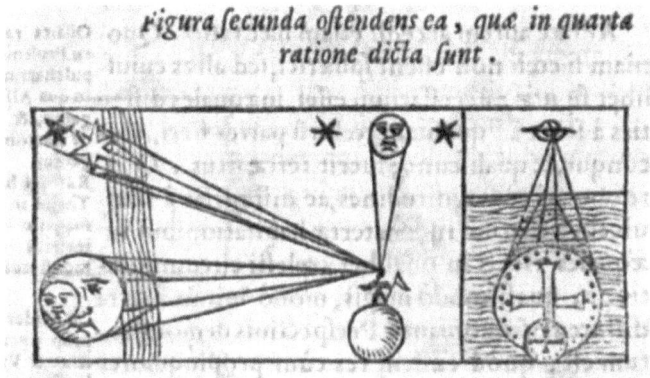

Fig. 4.2 Illustration from Barozzi's *Cosmographia* (Barozzi 1585, p. 18)

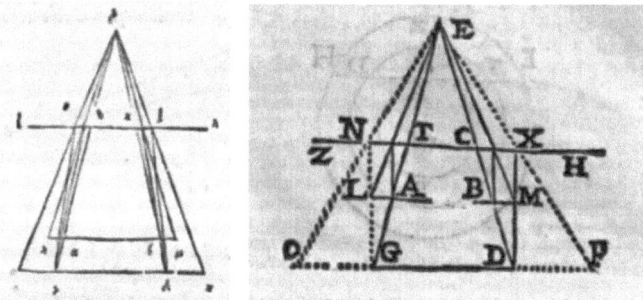

Fig. 4.3 The first diagram on the left is from the Greek *editio princeps* of the *Almagest* (Ptolemy & Theon 1538, p. 10 of Theon's commentary), while the second one on the right is from Della Porta's Latin translation of it (Ptolemy & Theon 1605, p. 19; Della Porta 2000, p. 31). According to Theon, drawing upon (pseudo-)Archimedes' work on catoptrics, when observed outside of water, the sizes of AB and GD are perceived to be identical because they are viewed under the same angle GED (see the diagram on the right). However, when these are submerged in water, with ZH representing the surface of the water, they appear larger. Due to refraction, AB will appear as LM, while GD as OP (see Smith 2003, pp. 104–107)

Expanding upon this demonstration, Theon offers a detailed explanation for the apparent magnification of celestial bodies when they approach the horizon. Assuming that vapors spread evenly above the Earth's surface up to a specific height, it can be inferred that these vapors create a spherical layer, centered on both the Earth and the universe. If an observer were located at the very center of the Earth, the distance to this vapor layer's edge would always remain constant. However, since an observer stands on the Earth's surface, their distance to the topmost boundary of vapors becomes greater when looking toward the horizon—specifically to the east and west—compared to when looking directly overhead at the zenith.[21] To clarify this concept, Theon provides an illustrative diagram with three concentric circles: the innermost circle represents the Earth; the middle circle denotes the vaporous sphere; and the outermost circle shows the path of the celestial body being observed (see Figs. 4.4, 4.5, 4.6. 4.7).

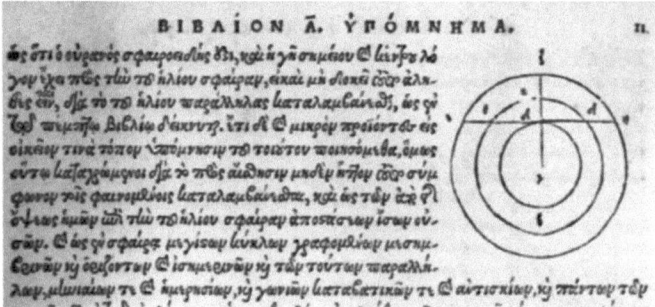

Fig. 4.4   Top portion of a page from the *editio princeps* of Theon's commentary on the *Almagest* (Ptolemy & Theon 1538, p. 11 of Theon's commentary)

between extramissionism and intromissionism, one can refer to Lindberg 1983, chapters I and IV. As far as I know, Galileo did not have a definitive view on this point. Sometimes he preferred to speak in terms of light rays, sometimes in terms of visual rays. For some example, see Galilei 2023, p. 86 n. 314.

[21] In Della Porta's Latin translation, the passage reads as follows: "Contingit autem, quamvis secundum omnes partes terrae facta exhalatione, ad orientales et occidentales maiores magnitudines astrorum videri, quod, in reliquis ex apparentibus circa ipsam – sphaerica ipsa et medio universi accepta –, consequatur magis oculis nostris extenso horizontis plano per maiorem humiditatem videri stellas" (Ptolemy & Theon 1605, p. 19; Della Porta 2000, p. 32, ll. 308–314).

Fig. 4.5 Diagram from Della Porta's Latin translation of Theon's commentary on the *Almagest* (Ptolemy & Theon 1605, p. 20; Della Porta 2000, p. 32)

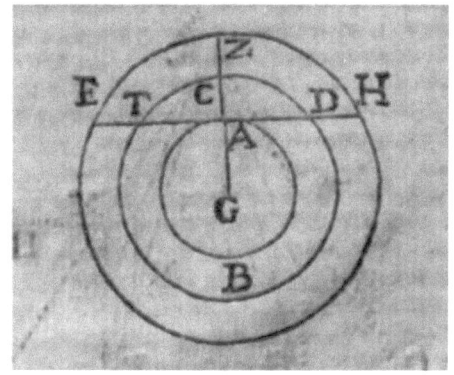

Theon contends that along the horizon, more vapor is interposed between the observer and the observed celestial body, as illustrated in the diagram from Della Porta's edition (see Fig. 4.5), where, with A being the location of the observer, AT equals AD, and both are greater than AC. This can be likened to an observer viewing an object submerged in deep water (AT and AD being "deeper" than AC).[22]

Thus, based on the Archimedean demonstration mentioned earlier, Theon clarifies why celestial bodies seem larger during their rise and set. This observation does not imply a non-spherical shape for the celestial vault and refutes the idea of it having a "lentil-like" (*lenticularis*) shape.[23]

---

[22] "Intelligatur terrae sphaera AB, circa centrum G, coelorum autem EZH. Amplius autem exhalatio non secundum omnes terrae partes similiter facta. Intelligatur rursus figura sphaerica, ut TCD, et per A habitationem producatur horizontis planum, ut faciat cum meridiano communem sectionem ETADH rectam, et producatur ipsi ad perpendicularem ab habitatione ad A linea ACZ, et producatur ad G centrum. Manifestum igitur, quod aequales sunt TA, AD invicem, et amplius utraque ipsarum maior AC, et semper linearum, quae sunt CT proprior AT, veluti horizontis maior longiori. Similiter autem et in CD, quae propter ipsam AD maior longiori, quare quando stella ad E, H apparebit, per longe maiorem humiditatem videbitur, quam quando in remotioribus, et super terram, ob id in horizontibus maiores, ut diximus, magnitudines astrorum videntur" (Ptolemy & Theon 1605, p. 20; Della Porta 2000, pp. 32–33, ll. 315–327).

[23] "Si quis autem dicet, nequaquam ex exhalatione in horizontibus maiores magnitudines videri astrorum, sed *lenticularem* supponens figuram coeli, ut minores quidem distantiae ad orientem et occidentem sint conversae, maiores autem ad meridianum, dixerit consistere quidem corpus coeleste, et stellas autem ferri, et propterea in horizontibus maiores magnitudines videri, in meridiano autem minores, falsa supponere arguetur,

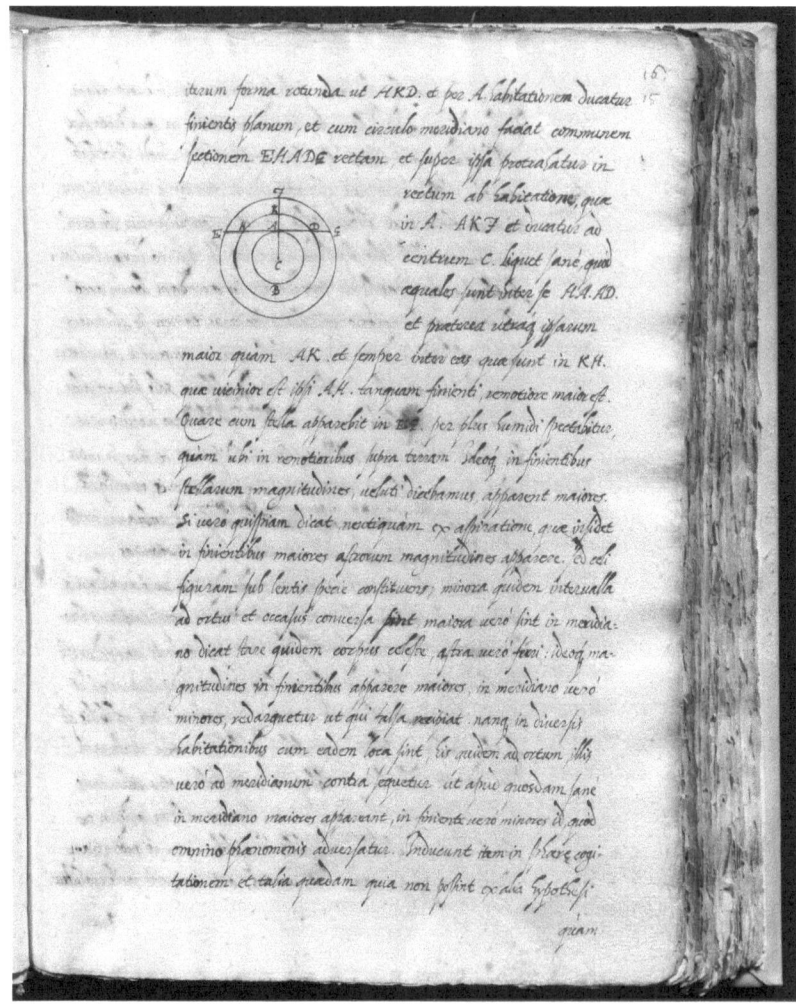

Fig. 4.6   Florence, Biblioteca Medicea Laurenziana, Ms. Acq. e doni 694, f. 15r. Used with permission from the Italian Ministry of Culture. Further reproduction by any means is prohibited

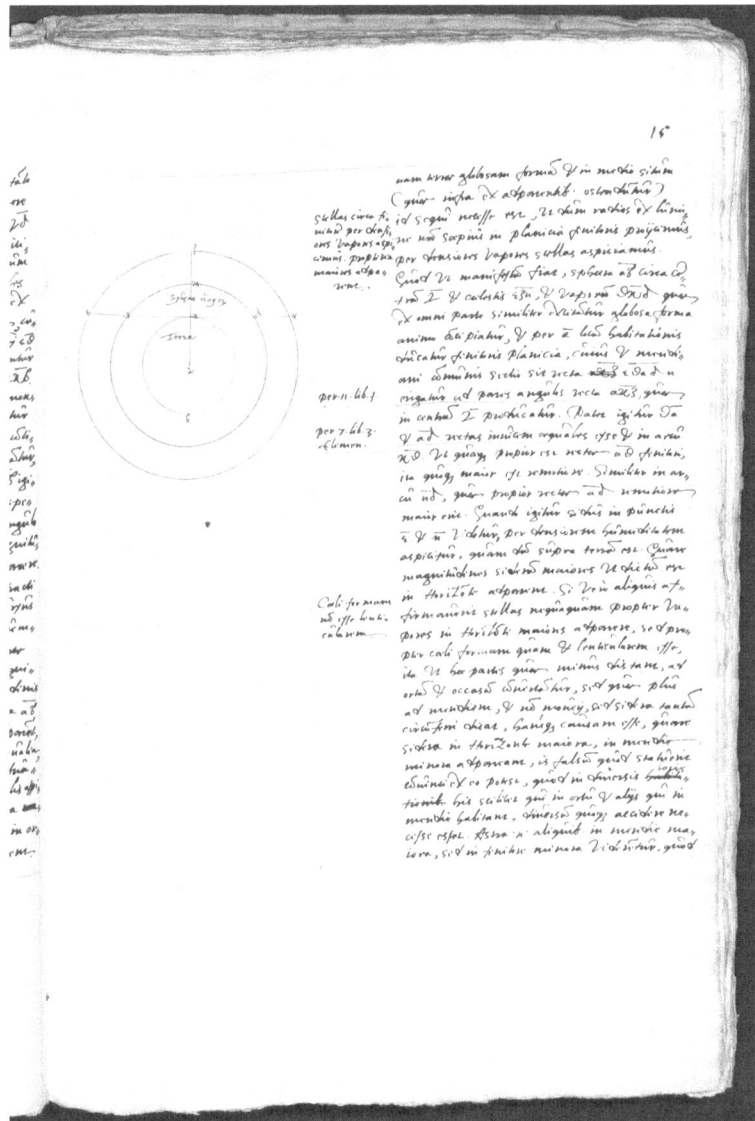

Fig. 4.7   Munich, Bayerische Staatsbibliothek, Clm 719, f. 15r. (Courtesy of the BSB)

GALILEO'S CRITICAL APPROACH TO *ALM.* I.3

The interpretation commonly accepted differed notably from the explanation given by Theon. As mentioned above, the more common understanding attributed to Ptolemy the idea that just above the Earth's surface, there existed distinct, relatively low layers, or walls, of vapors. These vapors, it was believed, distorted the observer's view, making celestial bodies seem to have a larger apparent diameter near the horizon, primarily due to refraction. On the other hand, Theon's interpretation extracts from the *Almagest* an atmospheric theory (that is, a theory based on the spherical shape of vapors surrounding the Earth) that, by analogy with the observations of objects in water, offers a geometric rationale for the magnified appearance of celestial bodies near the horizon.[24]

As for Galileo's interpretation, from the evidence present in *De motu antiquiora*, it is clear that he critically read the passage from *Alm.* I.3. One can also make educated guesses about his potential commentary. Galileo would likely have turned to empirical evidence to debunk widely held misconceptions. For instance, the belief that an object looks larger from the outside when submerged deeper in water.[25] He would have emphasized that the water, or the medium between the observer and

---

quod in differentibus habitationibus eorundem locorum, his quidem ad orientem existentium, his vero ad meridianum consequatur oppositum oportet, apud aliquos quidem ad meridianum maiora astra videri, ad horizontem vero minora, quod omnino adversatur apparentiis" (Ptolemy & Theon 1605, p. 20; Della Porta 2000, p. 33, ll. 327–338, emphasis added).

[24] Barozzi, too, recognized that Theon's explanation was a "geometric" one. However, he did not include it in his *Cosmographia*. See Barozzi 1585, p. 17 in the right margin: "Rationem hanc [i.e. quartam] Theon in Almagesto Geometrice demonstravit"; see also *supra*, n. 7 in this chapter. Prior to Barozzi, Piervincenzo Danti, in his vernacular annotations to the *Sphere* (a book owned by Galileo; see Favaro 1886, p. 252), made reference to Theon's geometric demonstration: "l'apparenza della disgregatione de i raggi visivi, che fa il corpo diafano, è da Teone dimostrata geometricamente nel secondo capitolo del primo libro dell'Almagesto" (Danti 1571, p. 11). Here Danti is evidently recalling Theon's reference to Archimedes.

[25] Galileo's appeal to experience is not entirely unproblematic and deserves thorough study, akin to what Thomas Settle did many years ago for another experience reported in *De motu antiquiora* (see Settle 1983, pp. 12–14). Here, it suffices to stress that Galileo sought a rational explanation for why objects underwater appear bigger, but his quest was in vain. This failure may imply that Galileo was dissatisfied with the demonstrations he knew from other sources, prompting him to develop his own in accordance with experience. See quotation *infra*, n. 28 in this chapter.

the observed, is not the primary cause of this perceived magnification. According to Galileo, simply having a greater quantity of water does not necessarily lead to increased optical enlargement. He believed that the main factor behind optical magnification was the shape or form of the medium.[26]

Before delving into Galileo's passage in its full detail, it is crucial to understand its context within the *De motu antiquiora* treatise. The passage is located in a chapter titled *A quo moveantur proiecta*, translated as *By what projectiles are moved*. In this chapter, Galileo challenges the widely accepted belief—promoted by Aristotle and his followers—that the medium is responsible for imparting motion to a projectile.[27]

As the chapter concludes, Galileo also takes a moment to critique another widely held belief related to the medium, one that had been accepted without thorough examination. In the following citation, I have enclosed Galileo's marginal addition within double square brackets:

> Such is the belief regarding objects submerged in water, which common opinion asserts appear larger than they truly are. [[And there is another common error of some who claim that, no matter how small, each particle of the mirror reproduces the entire image.]] However, since I could not find a cause for this effect, I finally turned to experience and discovered that a coin submerged deep in water in no way appears larger, but rather smaller. Therefore, I suspect that the person who first put forth this idea

---

[26] Della Porta, in his *De refractione* (1593), had also focused on the shape of the vessel: "But there is also another explanation, *based on the round form of the vessel*, for why things appear magnified under water and through glass. We will delve into this explanation when we discuss the glass sphere" ("Sed cur sub vitro et aquis maiora videantur, aliam quoque habet rationem *ex rotunda vasis forma*, quam reddemus quum de vitrea pila loquemur"; Smith 2017, p. 38 (Latin), 39 (English, slightly modified), emphasis mine). The end of this passage has been linked, by both Riccardo Bellé and A. Mark Smith, to the second book of *De refractione* (see Bellé 2004, p. 110, n. XI; Smith 2017, p. 39). While this link is accurate, I have not yet identified the precise section containing the explanation referred to by Della Porta. Nevertheless, Della Porta's reference to the round form of the vessel is noteworthy, especially when compared to Galileo's similar assertion in *De motu antiquiora*. However, this is obviously not sufficient to place Della Porta's *De refractione* among the sources of *De motu antiquiora*. Raffaello Caverni was the first to recognize this strong similarity between *De motu antiquiora* and Della Porta's passage just quoted (see Caverni 1891–1900, vol. I, p. 347). He read the passage as though Della Porta's opinion was taken from Seneca. Accordingly, he traced Galileo's argument in *De motu antiquiora* back to Seneca's. But this connection is not straightforward.

[27] See EN, I, pp. 307–314.

was led to this opinion during the summer, when sometimes a plum or others type of fruit are placed in a glass cup full of water, the shape of which resembles the surface of a cone; to those observing in such a way that the rays pass through the glass, these fruits seem much larger than they actually are. Yet it is not the water, but the shape of the cup, that causes such an effect, as we have explained in more detail in our commentary on Ptolemy's *Great Construction*, which (God willing) will be published soon. A sign of this is that, with the eye placed above the water in such a way that it can look at the plum without the intervening medium of the glass, it does not appear larger.[28]

It must be noted that reflection and refraction were traditionally explained by analogy with the motion of projectiles in different media.[29] So, it is probably not by chance that Galileo mentions common mistakes in catoptrics and dioptrics in a chapter of *De motu antiquiora* dedicated to the motion of projectiles.[30]

Galileo challenges two opinions which were indeed quite common at the time. The one added in margin, that every single part of the mirror reflects the whole image, can be found in Alhacen and was held by Guidobaldo del Monte in his theorem on mirror, which was never printed. According to Guidobaldo, "the species of the image is received in the mirror at every point of the mirror" (... *imaginis speciem in speculo in omnibus speculi punctis totam recipi*). He contended that the species of

---

[28] "Tale quiddam est quod creditur de rebus sub aqua existentibus, quas communis opinio asserit maiores, quam vere sint, apparere. [[Et est alius error communis quorumdam qui dicunt, quamlibet speculi particulam totam imaginem repraesentare.]] Cum autem talis effectus causam invenire non possem, tandem, ad experientiam accedens, inveni, nullo modo denarium in aquae profundo manens maius apparere, sed potius minus: quare arbitror ego, eum, qui primus hanc protulit sententiam, in hanc deductum fuisse opinionem aestivo tempore, cum interdum pruna vel alii fructus in vitreo calice aqua pleno, cuius figura conoidis superficiem referat, imponuntur; quae quidem, aspicientibus ita ut radii per vitrum transmittantur, longe maiora quam sint appareant. Verum non aqua, sed calicis figura, talis effectus causa; ut fusius in commentariis super Magnam Ptolemaci Constructionem declaravimus, quae (Deo favente) brevi eduntur. Signum autem huius est, quod oculo super aquam posito, ita ut non intercedente medio vitro prunum intueri queat, non maius apparet" (EN, I, p. 314).

[29] On such mechanical analogy, see Lindberg 1983, pp. 23–38. See also Dupré 2002, pp. 32–33, and the literature cited therein for more insights.

[30] On the contrary, Drake argued that in *De motu antiquiora*, Galileo "turned to an optical phenomenon *unconnected with motion in any way*" (Drake 1987, p. 103, emphasis added).

the image is received in "whatever point of the mirror" (*in quolibet puncto speculi*).[31] It is plausible that Galileo acquired knowledge of this theorem directly from Guidobaldo through an exchange of letters, now regrettably lost, or conceivably during a personal encounter in 1592.[32] However, the characterization of it as a "common error" (*error communis*) by Galileo does not conclusively designate Guidobaldo as his primary source. Indeed, a similar observation was articulated by Sarpi in the 1570s, likely based on Alhacen and Witelo.[33] In one of his personal notes, Sarpi wrote:

> Reflection is made from every point of the mirror, but from only one point with regard to one eye; and for this reason, only one [image] is seen on a plane mirror. In concave mirrors, more than one image can be seen because reflection can occur from multiple points [of the mirror] with regard to one eye.[34]

From this, one may argue that in the last decade of the sixteenth century Galileo was probably aware of the commonly held views on reflection espoused by Alhacen and Witelo, as presented in the *Opticae thesaurus* (1572), a text housed in his library.[35] He seemed to be critical of such tradition, but it is difficult to guess why.[36]

More interestingly, Galileo challenges another widely held belief, which finds mention in *Alm.* I.3. He delves into understanding how such a misconception took root, tracing it back to long-established cultural traditions. He firmly posits that it is the shape or form of the medium, rather

---

[31] Cited in Dupré 2002, p. 229 n. 153 (where, unfortunately, the abbreviations are not spelled out). According to Dupré (see ibid., n. 152), Guidobaldo has in mind Alhacen's *De aspectibus*, book IV, prop. 21 (see Alhacen & Witelo 1572, *De aspectibus*, p. 114).

[32] On the 1592 encounter, see Renn 1992, pp. 138–164, and Renn et al. 2001, pp. 79–89.

[33] According to Luisa Cozzi (reference in the following note), Sarpi referred to Alhacen's *De aspectibus*, book IV, props. 21–22, and book V, prop. 24. She also singled out the corresponding props. 24, 30, 45, in Witelo's *Perspectiva*, book V: see Alhacen & Witelo 1572, pp. 114, 133 of *De aspectibus*, and pp. 202, 205, 210 of *Perspectiva*.

[34] "Da ogni punto dello specchio si fa la riflessione, ma rispetto ad un occhio da un solo, e da qui nasce che nelli piani si vede una sola imagine: ne' cavi se ne può veder più d'una, perché da più punti rispetto ad un occhio la riflessione far si può" (Sarpi 1996, p. 75, thought n° 60).

[35] See *supra*, n. 8.

[36] It is worth noting that Galileo uses a more physical-oriented terminology; he prefers *particula* ("particle") over *punctum*.

than the medium itself, that primarily influences the optical magnification effect. He also hints at his commentary on the *Almagest*, indicating he has elaborated on this subject in more detail there. By so doing, he implies that the apparent magnification of celestial bodies near the horizon can be explained based on the premises he briefly outlines: (*a*) the amount of vapors positioned between the observer and the observed celestial body is not the cause of optical magnification; (*b*) the external shape of these vapors is the primary cause, instead.

To fully comprehend his stance, it would be essential to reference other works where Galileo delves into the same topic. Fortunately, such writings exist. For instance, in *The Assayer*, while debating Orazio Grassi's claims, Galileo argues that,

> not because of the light of vapors, but *because of the spherical shape of their outer surface, and due to the greater distance of that from our eye when the objects are closer to the horizon*, these objects [viz. the Sun and the Moon] appear larger than their common apparent size, and not just the luminous ones, but any other placed outside of such a region [of vapors].[37]

In both *De motu antiquiora* and this passage, Galileo underscores the importance of the "spherical shape" of the vapors' external surface when trying to comprehend the cause of optical magnification near the horizon. However, he introduces an added nuance not discernible from *De motu antiquiora*. The extent of the optical magnification hinges on the distance between the observer and the vapors' external surface. A larger distance amplifies the magnification effect. Rather than offering a geometric optical proof to elucidate this idea, Galileo prompts the reader to verify it empirically using a lens:

---

[37] "[...] non pel lume de' vapori, ma *per la figura sferica dell'esterna loro superficie, e per la lontananza maggiore di quella dall'occhio nostro quando gli oggetti son più verso l'orizzonte*, appariscono essi oggetti maggiori della lor commune apparente grandezza, e non i luminosi solamente, ma qualunque altro posto fuor di tal regione" (EN, VI, p. 354, emphasis added).

Place a convex crystalline lens between your eye and any object at various distances: you will see that when this lens is closer to the eye, the appearance of the viewed object will increase only slightly; but as you move it away, you will subsequently see it enlarging.[38]

Galileo replaces Ptolemy's comparison of objects submerged in water with the experience of using a lens. In this hands-on approach, in line with the arguments from *De motu antiquiora*, he shifts attention away from the medium between the observer and the observed. Instead, he emphasizes the significance of the distance between the observer and the "convex crystalline lens." As a result, since

the vaporous region ends in a spherical surface, not very high above the convexity of the Earth, the straight lines drawn from our eye to the said surface are unequal, and the smallest of all is the perpendicular towards the vertex, and of the others, the more inclined ones towards the horizon are larger than those towards the zenith.[39]

Galileo presents an atmospheric rationale for the phenomenon, resembling Theon's approach in his commentary on the *Almagest*.[40] Like Theon, Galileo emphasizes the importance of the distance between the observer and the atmospheric surface. However, their interpretations

---

[38] "Traponete tra l'occhio vostro e qualsivoglia oggetto una lente convessa cristallina in varie lontananze: vedrete che quando essa lente sarà più vicina all'occhio, poco si accrescerà la specie dell'oggetto veduto; ma discostandola, vedrete successivamente andar quella ingrandendosi" (ibid.).

[39] "[…] la region vaporosa termina in una superficie sferica, non molto elevata sopra il convesso della Terra, le linee rette che tirate dall'occhio nostro arrivano alla detta superficie, sono disuguali, e minima di tutte la perpendicolare verso il vertice, e dell'altre di mano in mano maggior sono le più inclinate verso l'orizzonte che verso il zenit" (ibid.).

[40] That it is an atmospheric rationale is also clear from what Galileo adds immediately after: "Quindi anco (e sia detto per transito) si può facilmente raccorre la causa dell'apparente figura ovata del Sole e della Luna presso all'orizzonte, considerando la gran lontananza dell'occhio nostro dal centro della Terra, ch'è lo stesso che quello della sfera vaporosa […]." This was indeed a phenomenon usually explained (for example, by Christoph Scheiner) via atmospheric refraction (see Galileo 2023, pp. 262–263, n. 953). Note that in the *Astronomic operations* Galileo claims that this phenomenon (i.e., the Sun's elliptical shape near the horizon) does not happen always. For this reason, he posits a distinction between two kinds of refractions due to (1) the atmosphere and (2) other denser and lower vapors, and attributes the phenomenon to the latter (see EN, VIII, p. 462).

diverge at a key point: Theon suggests that a greater distance equates to a higher volume of vapors, leading to enhanced optical enlargement. Galileo, on the other hand, believes that the sheer volume of the vaporous medium does not induce any perceptible magnification. He concentrates primarily on the distance between the observer and the convex-concave surface of a denser medium.

THEON'S COMMENTARY AND GALILEO'S STUDY OF THE *ALMAGEST*

The similarities and differences between Theon's and Galileo's explanations are evident in another text as well. Some scholars have remarked upon the parallels between Galileo's explanation and the one presented by Alimberto Mauri—likely a pseudonym—in the *Considerations*, published in 1606. This book emerged as a response to another one by Ludovico Delle Colombe. The controversy revolved around the *nova* of 1604.[41]

In consideration 43, Mauri, too, invokes the imagery of a "concave-shaped crystal" filled with "water or another medium" whose surface is thus convex. Before this, he posits that vapors form a sphere above the Earth's surface, reaching a height of up to 52,000 paces.[42] This interpretation aligns with Galileo's in *The Assayer*, but Mauri further supports his stance with a visual representation (see Fig. 4.8):

> Therefore, because that concave surface, through which our visual rays pass when we gaze at the Sun located in the east and in the west, is farther from us than the surface through which they pass when we gaze at it at midday; it is no wonder, I would say, that it appears larger to us both at sunrise and at sunset. This is because, to see it in any other position, our line of

---

[41] Matteo Cosci is working on a critical edition of this text, which was not included in the National Edition of Galileo's works. Favaro accorded credibility to a missive from which one may deduce that Galileo affirmed to Ludovico Delle Colombe his non-authorship of the *Considerations* (see EN, X, *Ludovico Delle Colombe to Galileo*, pp. 176–177). An English rendition, regrettably not consistently faithful, was curated by Drake in Galilei 1976, pp. 73–130.

[42] This unit circulated widely in the late-Renaissance and was taken from pseudo-Alhacen's (i.e., Ibn Mu'adh's) *De crepusculis et nubium ascensionibus* (see Alhacen & Witelo 1572, *De crepusculis*, p. 287; Sabra 1967; Goldstein 1977; Smith 1992; Goldstein & Smith 1993; Smith 2003, pp. 112–114). The same unit was also mentioned in Witelo's *Perspectiva* (see Alhacen & Witelo 1572, *Perspectiva*, book X, prop. 60, p. 453).

Fig. 4.8 Diagram from Mauri's *Considerations* (Mauri 1606, f. 24v)

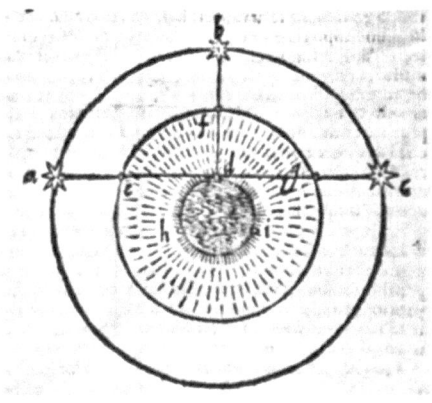

sight must pass through a point closer to us, as can be clearly seen in the current figure.[43]

In the recent commented edition of *The Assayer*, edited by Michele Camerota and Franco Giudice, this illustration is rightly reproduced to provide clarity to Galileo's discourse.[44] This segment of Mauri's *Considerations* had already been attributed to Galileo and linked to *The Assayer* by Francesco Piccolino and Nicholas J. Wade.[45] Additionally, Noel Swerdlow, in a yet-to-be-published work, noticed substantial similarities between the aforementioned segment from *De motu antiquiora* and Mauri's exposition.[46] Given Galileo's notably unique rationale for

---

[43] "Perché adunque è più distante da noi quella concava superficie, per la quale passano i nostri raggi visuali, quando rimiriamo il Sole posto nell'Oriente, e nell'Occidente, che quella, per la quale passano, quando lo rimiriamo nel mezzogiorno; non è maraviglia, direi io, ci appaia ei maggiore, e nel nascere, e nel tramontare, poiché, per vederlo, in qualsivoglia altro luogo, dobbiamo passare con la veduta a noi più vicina, si come manifestamente si scorge nella presente figura" (Mauri 1606, f. 24v).

[44] See Galilei 2023, p. 262 n. 951.

[45] See Piccolino & Wade 2013, pp. 231–235.

[46] Matteo Cosci and I both independently made the same observation, unaware of each other's findings. Still, the credit ultimately goes to Swerdlow. Cosci—to whom I am grateful—informed me of Swerdlow's observations by email on 5 December 2022. The day after he kindly sent to me a short extract of Swerldow's unpublished work. Here, Swerdlow holds that in the *Considerations* "Galileo makes the same point as in *De motu* to explain the larger apparent size, that it is not the density of the water or the crystal

the perceived magnification of celestial bodies near the horizon—clearly diverging from prevalent theories of the time—I believe it acts as a hallmark of Galileo's direct authorship in Mauri's *Considerations*.

While the idea of attributing part or even the whole of the *Considerations* to Galileo is not new,[47] what remains unobserved is that, if Galileo wrote under the pseudonym Mauri, he used a figure that mirrors the one employed by Theon in his commentary on *Alm.* I.3 (cf. Figures 4.4, 4.5, 4.6, 4.7, 4.8).

What is particularly intriguing is that Galileo reemploys this same diagram but associates it with an interpretation that is distinct from Theon's.[48]

I believe that Galileo might have previously formulated an explanation—before or at the same time as the writing of *De motu antiquiora*—regarding the optical magnification of stars near the horizon, which bears a close resemblance to the perspectives shared in both *Considerations* and *The Assayer*. However, it remains uncertain whether Galileo, during his stay in Pisa, had the opportunity to peruse Theon's commentary on the *Almagest*. It is worth noting that the *Considerations* dates back to 1606; just a year before, Della Porta's Latin translation of the first book of the *Almagest*, complemented by Theon's commentary, saw its publication in Naples.

Based on the research so far, I believe it is possible that Galileo referred to Theon's work when he studied, and perhaps even commented on, the

---

but the convex shape the water is given by the glass or decanter that makes the coin or fruit appear larger, and the same applies to a candle flame behind the decanter. [...] And continuing the argument of *De motu*, this also shows the error of those who say that vapors near the horizon make the sun appear larger by providing a denser path through which our sight passes, separating our visual rays as does water, which is denser than air, and making objects appear larger." Swerdlow considered Mauri's consideration 43 "to show that it carries over an argument from *De motu*, as evidence for Galileo's authorship of the *Considerations*" (Swerdlow 2021, 5, pp. 41–42). I hope this posthumous book will soon see the light of day.

[47] Drake had previously credited Galileo with authoring the *Considerations* in Galilei 1976.

[48] As brought to my attention by Michele Camerota, Galileo himself claims in *The Assayer* that geometric diagrams, in reality, provide limited insight: "in all the works of mathematicians, little consideration should be given to diagrams [*figure*] whenever there is written explanation [*scrittura che parla*]" ("... in tutti i libri de' matematici niun riguardo si ha già mai delle figure, tutta volta che vi è la scrittura che parla"; EN, VI, p. 306).

Almagest. I contend that this possibility is stronger than the suggestion made by Piccolino and Wade. They suggested that Galileo drew inspiration from certain manuscripts of Leonardo da Vinci, which were possibly owned by Gian Vincenzo Pinelli and preserved in his Paduan library.[49] Yet the two scholars overlooked *De motu antiquiora*. In this work, as has been showcased, Galileo already approaches, albeit indirectly, the subject of the apparent magnification of celestial bodies, given his reference to the *Almagest*. I harbor reservations about the idea that Galileo could have easily accessed Leonardo's manuscripts during his time in Pisa.

If one then assumes that Theon's commentary was a source of Galileo, one must also acknowledge the discerning mindset with which he approached it. Galileo did not blindly adopt Theon's reconciliatory views toward Aristotle's natural philosophy and Ptolemaic astronomy. He also did not find Theon's reference to Archimedes persuasive in the debate surrounding the apparent enlargement of celestial bodies. As such, the impact of Theon's commentary on Galileo's understanding of the *Almagest* warrants a thorough investigation. Recognizing its potential importance to Galileo is merely an initial yet captivating step in this exploration.

Conclusion

While it cannot be definitively proven that Galileo studied the *Almagest* using Theon's commentary, it seems reasonable to believe he perused it at some point. This is especially likely if we accept the view that the *Considerations* by Alimberto Mauri were, at least in part, written by Galileo. In that text, Mauri employs a diagram identical to Theon's from his *Almagest* commentary. This diagram is used to illustrate a unique theory of Galileo, different from that of Theon and others supported at the time, and in line with some general premises already outlined in *De motu antiquiora*. These revelations provide intriguing insights that, with further exploration, could deepen our understanding of how Galileo engaged with the *Almagest*.

In this chapter, I have examined a theory put forward by Galileo to elucidate the enlargement of the apparent diameter of celestial bodies near the horizon. It is crucial to underscore that this is just one among the

---

[49] See Piccolino & Wade 2013, pp. 208, 229.

theories posited by Galileo. In his *Sidereus Nuncius* (1610), Galileo introduces an alternative theory—one that can be considered more traditional from one standpoint, yet remarkably peculiar from another. Traditional in its assertion that the perceived enlargement of the Sun and the Moon results from the interposition of vapors between the observer and the two "luminaries" (*luminaria*). Peculiar, as Galileo further contends that this same theory explains the observed diminution in size of fixed stars and other planets.[50] The attempt by Isabelle Pantin, in her masterful edition of the *Sidereus Nuncius*, to reconcile this theory with that of the *Perspectivi* Alhacen and Witelo encounters certain challenges.[51] In this context, it may be useful to also consider the, albeit frequently biased, objections raised by Francesco Sizzi. In the *Dianoia astronomica* (1611), he criticized this passage of the *Sidereus Nuncius*, posing a rhetorical question:

> Who among astronomers and opticians has ever expressed such an opinion in his writings?[52]

Many years later, Benedetto Castelli would disclose another theory supported by Galileo. In 1639, he wrote the *Discourse on sight* (published posthumously in 1669), in which the so-called "Moon illusion" is addressed, among other topics, and is treated as "a fallacy of judgment and of apprehension" (*fallacia del giudizio, e dell'apprensione*). Consequently, he realized that the retinal image of the Big Dipper, along with

---

[50] "Constat, terrestium vaporum obiectu Solem Lunamque maiores, sed fixas atque Planetas minores, apparere: hinc Luminaria prope horizontem maiora, Stellae vero, minores ac plerunque incospicuae, imminuuntur etiam magis, si iidem vapores lumine fuerint perfusi; idcirco Stellae interdiu ac intra crepuscola admodum exiles apparent" (EN, III, *Sidereus Nuncius*, p. 95). Galileo may have been melding two theories in this context. The first is about the perceived magnification of the Sun and the Moon, attributed to the illumination of terrestrial vapors—a theory later subject to criticism in *The Assayer* (see EN, VI, pp 353–354). The second theory involves the so called "adventitious rays," from which it may be inferred that fixed stars appear larger in the absence of light but seem to diminish in size when subjected to surrounding illumination. However, this interpretation may be forced. The passage is not easily understandable.

[51] See Galilei 1992, pp. 93–94 n. 168. Reading this note is worthwhile for gaining a clear understanding of the backdrop against which the passage from Galileo's *Sidereus Nuncius* should be interpreted.

[52] "Quis unquam astronomorum aut opticorum talem scriptis suis consignavit opinionem?" (EN, III, p. 242).

that of any celestial body, remains of constant size, despite our perceptual tendency to interpret it as larger when positioned at the horizon. The fallacy, in essence, does not reside in the eye but in our capacity to estimate magnitudes.

In short, Castelli believed that our estimation of the size of celestial objects is always formed by comparison with other intervening objects whose size serves as a point of reference. For instance, the Big Dipper appears larger when juxtaposed against a vast field than when measured against a smaller portion of a roof. The fallacy, or illusion, stems from the manner in which we process diverse visual data to gauge the dimensions of celestial bodies.[53]

Having shared this insight with Galileo, Castelli was informed by his mentor of "a much more subtle and artful illusion":

> I felt much satisfied by the thought above until on communicating it to my teacher he unveiled to me a much more subtle and artful illusion in which our judgement is tangled up and deluded. Although I do not have the ability [*animo*] to explain it with that felicity [of expression] by means of which it was clarified to me by that great man, as he is ever used to do in all his discourses about obscure and recondite matters of Nature, however difficult, I shall attempt to represent it in the best way possible to me while begging whoever reads to excuse me if I shall not be able to reproduce as clearly what I was then taught. Thus, I first consider that if two unequal objects placed at various distances are judged to be equal, it must be that the judgement about their sizes is fallacious. [...] Similarly, if two objects are truly equal and are truly placed at equal distances from our eye but one of them is judged by us to be farther away, it will be appraised larger. [...] In sum, in these operations of our judgement, to be deceived with respect to distances results in being deceived in our judgement of magnitudes, and also because of the latter we come to form a false judgement about the distance. But to go back to our purpose, when we lift our sight to the contemplation of the heavens and of those objects that are commonly seen in them we form a very false conception of their disposition. For we reckon those parts that are near the vertex as very near to the eye and those that are placed along the horizon as very far away.[54]

---

[53] See Castelli 1669, pp. 28–30; Castelli 2018, pp. 118–119.

[54] "Dal sudetto pensiero rimasi assai sodisfatto, e questo fintanto che comunicandolo con il mio Maestro mi fù da lui scoperto un altro inganno molto piu sottile, et artificioso, nel quale il nostro giudizio viene avviluppato, e deluso. E perche non mi dà l'animo di spiegarlo con quella felicità, che mi fù da quel grand'uomo dichiarato, come egli è solito

Galileo added a further nuance to Castelli's explanation: our judgment regarding the size of celestial objects is flawed because we mistakenly perceive them as being closer to us at the zenith or farther from us at the horizon, based on comparisons with other intervening objects. Consequently, the erroneous estimation of the size of celestial objects stems from a misjudgment of their distance.

As Cornelis Plug and Helen E. Ross pointed out in their historical review of the "Moon illusion," "there are two rival explanations for the effect of intervening objects – relative size and perceived distance."[55] Castelli reports both explanations in his *Discourse*: the relative-size explanation is ascribed by Castelli to himself, while the perceived-distance explanation is attributed to Galileo. They are not regarded as rival but rather as complementary.

When examining Galileo's various theories in light of later scientific discoveries, especially when viewed in chronological order, the resulting implications may appear "ironic." As Piccolino and Wade have suggested, the "rather ironic conclusion to the story is that near the end of his life Galileo, contrary to his earlier views, gives an accurate explanation of the visual processes involved in the perception of form and distance and thus of the celestial illusions."[56]

---

sempre fare in tutti i suoi discorsi, ancorche difficilissimi, & intorno a materie oscure, e recondite della Natura, per tanto procurerò rappresentarlo nel miglior modo a me possibile, pregando chi legge a scusarmi, se non saprò così vivamente replicare quanto mi fù allora insegnato. Prima dunque considero, che se due oggetti ineguali saranno collocati in varie lontananze siano giudicate eguali, seguirà che ancora il giudizio intorno alle grandezze di quegli oggetti sia fallace [...]. Similmente se due oggetti saranno eguali realmente, e realmente posti in distanza eguali al nostro occhio, ma uno di essi venga da noi giudicato piu lontano sarà stimato maggiore [...]. Et in somma in queste operazioni del nostro giudizio, se noi ci inganniamo nelle lontananze, ne siegue ancora l'inganno, nel giudicare della grandezza, dal che poi venghiamo ancora a formare falso giudizio della lontananza. Ora nel proposito nostro, quando noi solleviamo la vista alla contemplazione del Cielo, e di quegli oggetti, che in essi si veggono comunemente formiamo un concetto falsissimo della disposizione del Cielo, imperoche le parti sopra il nostro vertice ce le figuriamo assai vicine all'occhio, e quelle che sono collocate lungo l'orizzonte le apprendiamo molto lontane" (Castelli 1669, pp. 31–32; Castelli 2018, pp. 119–120). English translation from Ariotti 1973, pp. 16–17. I have corrected recurrent typos such as "father" instead of "farther." In his edition, Salvatore Ricciardo refers to EN, VIII, p. 626 after noting that the "master" mentioned by Castelli is evidently Galileo (see Castelli 2018, p. 120 n. 15).

[55] Plug & Ross 1989, p. 18.

[56] Piccolino & Wade 2013, p. 242.

However, it is imperative to acknowledge that in Galileo's time, explanations rooted in optics and perception theory were not always seen as inherently contradictory.[57] Therefore, it is not certain that in 1639 Galileo had abandoned the theory presented sixteen years earlier in *The Assayer*. Moreover, in 1651, Riccioli maintained that the apparent enlargement of celestial bodies near the horizon was not due to a perceptual error.[58] Even within the *Accademia del Cimento*, the issue was still being discussed using experiments and theories akin to those proposed in *De motu antiquiora* and *Considerations*.[59] In short, it does not matter

[57] See *supra*, n. 11 in this chapter.

[58] Riccioli thought to have experimentally confirmed it: "Respondi olim absolute videri maiora sidera in Horizonte, vi refractionis secundae a perpendiculari factae ob vapores aqueos inter aetheream auram, atmosphaerae nostrae exteriorem, et inter aërem oculo nostro propiorem, nec ita densum, ut sunt vapores illi horizontales; idque ita semper evenire, ut experimento patebit. Neque vero id tribuendum fallaciae aestimationis, ob interiectum ractum vallium, camporum, marium: nam si in cubiculo aut horto cernas Solem orientem vel occidentem, ita ut sepes, aut murus, aut margo inferior fenestrae prohibeant prospectum omnem interiacentis spatij usque ad Horizontem, nec aliud quam Solem videas, illud tamen enormiter amplioris videbis, quam longe ab Horizonte. Sed neque id tribuendum est merae dilatationi pupillae; nam si coneris eam voluntario motu constringere, adhuc tamen orientis occidentisve Solis, plenaeve Lunae imaginem multo maiorem videbis, quam alias, et idem in Iove, Venere, coeterisque siderisbus spectabis" (Riccioli 1551, p. 644, col. 1).

[59] See the letter transcribed and printed in Targioni Tozzetti 1780, vol. II, part 2, pp. 750–753. Raffaello Caverni attributed it to Leopoldo de' Medici. According to Caverni, Leopoldo composed this letter in response to another missive from Vincenzo Renieri. While adhering to Caverni's ascriptions, it is crucial to underscore that his interpretation of the letter's contents proves to be misleading. Caverni posited that Leopoldo was disputing the theory advanced by Renieri, aiming to elucidate the apparent augmentation of the apparent diameter of celestial bodies in proximity to the horizon. Also, Caverni attributed Renieri's theory to Galileo himself, leading to the inference that Leopoldo was refuting Galileo's explanation from *The Assayer*. Contrary to this interpretation, an analysis of Leopoldo's letter reveals Renieri's endeavor to expound the "Moon illusion" in analogy with the functionality of the telescope (see ibid., p. 751). But Galileo never employed the telescope as an example to explicate the apparent enlargement of celestial bodies near the horizon. In *The Assayer*, as already noted, he relies on the experience of a convex lens positioned at varying distances from the observer. Moreover, Leopoldo appears to endorse a view consistent with that of Galileo, asserting that the denser medium is not the cause of any optical magnification. To elucidate this phenomenon, Leopoldo, akin to Galileo and Mauri, resorts to an experiment involving a concave vessel partially filled with water, thus forming a rudimentary lens. This experiment serves to underscore the primacy of the lens surface's shape, dismissing considerations of its "thickness" (*grossezza*). Leopoldo concludes by mentioning that "another person" had conducted this experiment and drawn certain conclusions from it ("Questa Esperienza della Lente dell'Acqua, diede

that Gemma Frisius, Paolo Sarpi, and Johannes Kepler had already emphasized that it was solely a perceptual error.[60] Thus, it is extremely difficult
to establish the evolution of Galileo's theories on this matter.

At any rate, a discernible facet is Galileo's early skepticism, in *De motu
antiquiora*, toward the prevailing explanation—namely, the theory of a
wall of denser vapors. From what we can piece together, he advocated for
an atmospheric explanation of the phenomenon of apparent enlargement.
In doing so, he distanced himself from more or less mainstream views
linked to *Alm.* I.3.

occasione ad una Persona di considerare un'altro [*sic*] particolare"). The referent of this
statement remains somewhat ambiguous (was it Galileo?). However, it is unequivocal that,
upon meticulous textual scrutiny, Caverni's interpretation lacks substantive support (see
Caverni 1891–1900, vol. II, pp. 90–91).

[60] See Piccolino & Wade 2013, pp. 227–231.

CHAPTER 5

Concluding Remarks

Abstract The final chapter summarizes the key findings of the study. It highlights the alignment between Galileo's intention to write a commentary on the *Almagest* and the educational needs of his time, emphasizing Theon's commentary as essential for understanding Ptolemy. Galileo aimed to present himself as a supporter of Ptolemy, distinguishing Ptolemaic views from Aristotelian ones. However, examining *De motu antiquiora* also reveals challenges in fully understanding early Galileo's cosmology and his interpretation of Ptolemaic astronomy.

Keywords Galileo Galilei · Claudius Ptolemy · Theon of Alexandira · *Almagest* · *De motu antiquiora* · Hipparchus

What has this extended exercise in contextualization achieved so far? Nothing definite, certainly. However, it has undeniably enriched our understanding of how the *Almagest* was meant to be read and studied during the time when Galileo wrote *De motu antiquiora*. A certain coherence has emerged between Galileo's presumed intent to compose a commentary on the *Almagest* and the didactic needs of his time. It has been observed that within this framework, Theon's commentary was

I. Malara, *Galileo and the* Almagest, *c.1589–1592*, Palgrave Studies in the History of Science and Technology, https://doi.org/10.1007/978-3-031-70614-1_5

deemed by many as essential for tackling the challenging study of the *Almagest*.

Although this commentary currently does not hold a favorable reputation, it is apparent that, between the late sixteenth and the early seventeenth centuries, it was still regarded as a work of considerable merit.[1] It would be worthwhile to explore its reception in these centuries more extensively, seeking to understand the extent of its impact on the astronomy and cosmology of that period.

While it cannot be definitively proven that Galileo studied the *Almagest* using Theon's commentary, it seems likely, especially if we accept that Alimberto Mauri's *Considerations* includes Galileo's work. This text features a diagram identical to one used by Theon, but it illustrates a theory unique to Galileo, differing from contemporary views and aligning with ideas from *De motu antiquiora*. These insights offer promising leads for further exploration of Galileo's engagement with the *Almagest*.

On this note, if my interpretation of *De motu antiquiora* is plausible, it is worth noting that Galileo initially intended to present himself to his readers as a non-Aristotelian Ptolemaic.[2] He distinguished Ptolemaic geocentric arguments from Aristotelian ones and aimed to enhance Ptolemy's geocentrism with a doctrine of natural motion that could be described as Archimedean. This reveals the significant influence both Ptolemy and Archimedes had on Galileo from the beginning. However,

---

[1] For the unfavorable assessment by contemporary scholars, see Neuguebauer 1975, pp. 5–6. According to Toomer, Theon's commentary "is for the most part a trivial exposition of Ptolemy's text, explaining obvious points at excessive length. Despite Theon's promise to improve over previous commentators on the *Almagest*, 'who claim that they will only omit the more obvious points, but in fact prove to have omitted the most difficult,' the commentary is open to precisely this criticism. It is never critical, merely exegetic. To the modern reader it is almost useless for understanding Ptolemy" (Toomer 1976, p. 321). However, this (anachronistic) point of view has been recently challenged (see Bernard 2014).

[2] That he intended to publish his earlier writings on motion, it is evident from at least two passages of *De motu antiquiora*. (1) "Erunt multi qui, *postquam mea scripta legerint*, non ad contemplandum utrum vera sint quae dixerim, mente convertent, sed solum ad disquirendum quomodo, vel iure vel iniura, rationes meas labefactare possent" (EN, I, p. 412, emphasis added); (2) "Sed, quia haec omnia, quae in superioribus his duobus capitibus tradita sunt, minus adhuc mathematice, et magis physice, declarari possunt, reducendo ea ad lancis considerationem, placuit in sequenti capite convenientiam explicare, quam mobilia haec naturalia cum bilancis ponderibus servant: *et hoc ad uberiorem eorum quae tradentur cognitionem, et ad exactiorem legentium cognitionem*" (ibid., p. 257, emphasis added). They are both recalled in Abbatouy 2000, p. 20.

it must be acknowledged that the current study has not contributed to shedding light on specific aspects of early Galileo's cosmology and his full understanding of Ptolemaic astronomy.

As for cosmology, it is worth noting that, according to Pietro Daniel Omodeo and Irina Tupikova,

> a cosmological perspective like that of Ptolemy virtually entails a reversal of Aristotelian physics, once the arguments for terrestrial centrality are demonstrated to be invalid from an astronomical perspective, as Copernicus demonstrated in the first book of *De revolutionibus*.[3]

In principle ("virtually"), this conclusion may appear cogent, but in practical terms—within a historical context—it remains a proposition demanding substantiation.

Galileo's *De motu antiquiora* reveal a distinctive scenario. Although Galileo accepts Ptolemy's arguments for the Earth's centrality, the cosmological perspective of the *Almagest* does not prevent him from rethinking, in a non-Aristotelian way, the cause or rationale governing the order of the world.

Historically, *De motu antiquiora* have prompted some scholars to posit that Galileo was "still a prisoner of the traditional cosmological framework" due to its adherence to the Aristotelian ontological hierarchy of the elements (earth, water, air, fire). However, a nuanced analysis contradicts this notion. *De motu antiquiora* do not embody an "ontocosmology"—a cosmology where the world's elemental order is determined by its ontological characteristics.[4] Galileo argues that all four elements share a common material basis and are inherently heavy. He reduces their essence to the ratio between matter and volume. Thus, an Archimedean rationale, based on hydrostatic principles explains why the Earth, not fire, occupies the central position in the world.

Although Galileo retains much Aristotelian terminology, he moves away from the notion that an element's motion toward its natural place is a teleological process ending in rest. As Maarten Van Dyck noted, in *De motu antiquiora*, Galileo is able to

---

[3] Omodeo & Tupikova 2016, p. 171.

[4] According to Maurice Clavelin, within *De motu antiquiora*, Galileo is "toujours prisonnier du cadre cosmologique traditionnel [...]" (Clavelin 2016, p. 90 n. 8). See also *supra*, Chapter 1, n. 66.

rethink the Aristotelian cosmos from a fundamentally different perspective, rather than doing justice to it. He retains some of its overall characteristics but fills it out completely anew from the inside by replacing qualitatively differentiated elements with homogeneously structured matter.[5]

Adding to the intrigue, Galileo appears to recognize that the cosmological perspective in the initial chapters of the *Almagest* does not fully align with the Aristotelian view.

In Aristotle's view, the element earth is at the center of the world because it is its natural place—an argument that, upon closer scrutiny, turns out to be a tautology. Ptolemy, on the other hand, employs a different argumentative approach, which Galileo noticed. The Alexandrian astronomer first establishes that the Earth is a sphere and that heavy bodies naturally move toward its center. He then uses astronomical observations to argue that the Earth must be at the center of the world. Consequently, since heavy bodies move toward the Earth's center, they also naturally move toward the center of the world, making any further inquiry into why this happens redundant.

Galileo's distinctive interpretation of the early sections of the *Almagest* lies in attributing to Ptolemy a doctrine of gravity that diverges from Aristotle's—a gravity not understood as a natural inclination toward the center of the world. While Galileo agrees with Ptolemy that heavy bodies naturally gravitate toward the Earth's center, he acknowledges that this does not explain why, assuming Ptolemy's proofs hold, the Earth as a whole occupies the central position in the world.

Despite later offering an Archimedean response to this issue, it is noteworthy that Galileo identifies in the *Almagest* a doctrine of gravity that could potentially align with Copernicus' view. According to Copernicus, gravity is the natural tendency of the parts of a planet or star to reunite with their whole, thereby forming a globe.[6]

At this point, one might speculate that Galileo's favorable view of Copernicus' *De revolutionibus* was influenced by his reading of the cosmological sections of the *Almagest*. It could also be suggested that Galileo, recognizing Ptolemy's fallacies through Copernicus, converted to

---

[5] Van Dyck 2006, p. 110.

[6] See Copernicus 2015, vol. II, p. 32, ll. 12–18. For an English translation, see Rosen 1992.

heliocentrism. However, *De motu antiquiora* reveals potential complexities in this narrative. Although Galileo had a good understanding of *De revolutionibus* when writing *De motu antiquiora*, it is unclear whether *De revolutionibus* significantly shaped his interpretation of the *Almagest* at that time. Currently, there is insufficient evidence to confirm any specific hypothesis.

In general, Galileo's precise cosmological views during his time in Pisa evade thorough examination. *De motu antiquiora* proves inadequate for understanding his conceptualization of the world's system during that period. His early writings on motion regrettably offer only elusive indications, leaving room for ambiguity.

Consider, for instance, the following preparatory note:

> We call local motion that in which the mobile's center of gravity is in motion; hence, we refrain from attributing local motions to celestial orbs, as their center of gravity – also the center of their magnitude – remains perpetually at rest.[7]

Alongside defining local motion through the concept of the center of gravity, Galileo introduces the idea of celestial orbs. These orbs are spherical and possess weight, inferred from their "center of gravity" (*centrum gravitatis*), which aligns with the "center of their magnitude" (*magnitudinis... centrum*). This center remains stationary.

One might then infer that the orbs Galileo refers to are likely solid and composed of the same matter as the elements. Given their weight, it follows that their center of gravity should align with the world's center, consistent with the Archimedean geocentric *ratio* presented in both the dialogue and treatise versions of *De motu antiquiora*. In essence, these orbs would be homocentric spheres.

However, a closer examination reveals that these hypotheses are weak due to the limited information available. It is already notable if, based on Galileo's early writings on motion, we can reconstruct even a small fragment of his engagement with the *Almagest*, particularly his initial interaction with Ptolemy's first book. *De motu antiquiora* do not provide

---

[7] "Motum localem appellamus illum, in quo mobilis centrum gravitatis movetur: quare caelestium orbium motus locales non dicemus, cum eorum centrum gravitatis, quod magnitudinis etiam centrum est, immobile semper maneat" (EN, I, p. 416).

insights into his exploration of the remaining books, highlighting a significant gap that future studies should address by examining Galileo's later works.

There is, however, a passage in the first treatise version of *De motu antiquiora* that suggests Galileo had already examined the other books of the *Almagest*. Where Galileo discusses free fall acceleration, there is another reference to Ptolemy. After outlining the acceleration theory mentioned in the introduction to this book, Galileo adds:

> when I had conceived this [theory about free-fall acceleration], and, two months later, happened to read what is written by Alexander on this matter, I understood from him that this was also the opinion of that most learned *philosopher*, namely Hipparchus, praised by a most learned man, Ptolemy, who highly esteemed him and extol with the highest praises throughout the *Almagest*.[8]

This passage is challenging because Galileo claims to have learned about Hipparchus' doctrine from Alexander of Aphrodisias, but by Galileo's time, that source was no longer available. Hipparchus' doctrine was known through Simplicius' commentary on *De Caelo*, which was printed in Latin several times in the late sixteenth century. However, Simplicius only cites Alexander's criticisms of Hipparchus. Thus, Galileo might not have learned about Hipparchus directly from Simplicius. Some scholars suggest that Galileo's source could have been Benet Perera's *De communibus omnium rerum naturalium principiis et affectionibus* (1576).[9]

At any rate, it is important to note that neither Simplicius nor Perera refers to Hipparchus as a "philosopher," unlike Galileo. Traditionally, Hipparchus was known for his contributions as an astronomer. The only exception is the *Suda*, a Byzantine encyclopedia from the tenth century,

---

[8] "Et hanc veram existimo causam accelerationis motus: quam quidem cum excogitassem, et, post duos menses, forte quae ab Alexandro de hac re scribuntur legerem, ex eo intellexi, hanc quoque fuisse sententiam doctissimi illius *philosophi* a doctissimo viro laudati, a Ptolemaeo, nempe, a quo magni habetur et summis laudibus per totum suae Magnae Constructionis contextum extollitur Hipparchus" (EN, I, p. 319, emphasis added).

[9] For an accurate overview on this matter, see Guerrini 2014.

which was widely circulated during the Renaissance.[10] The Adler edition of the *Suda* describes Hipparchus as "from Nicaea, a philosopher, lived at the time of the consuls."[11]

The only reference to the *Suda* made by Galileo is found in *The Assayer* and is evidently drawn, in a polemical manner, from Orazio Grassi's *Libra*.[12] Generally, it remains unclear how acquainted Galileo was with this encyclopedia. At present, I posit that the most plausible and economical hypothesis is that Galileo derived the designation "philosopher" for Hipparchus from Gerard of Cremona's translation of the *Almagest*.

In the Greek rendition, Hipparchus is repeatedly described as *philóponos* and *philalēthēs*, signifying, respectively, a "diligent" (literally, "lover of work") and a "lover of truth." George of Trebizond renders them as *industrius* and *amicus veritatis*, respectively. But Gerard of Cremona seems to translate from an Arabic version in which at least one instance of *philóponos* had been interpreted as *philosophos*, meaning philosopher or, literally, "lover of knowledge" (*amator scientiae*):

> The first thing we need to explain about the cause of the [motion of the] Sun is the discovery of how its annual longitude is determined, and [the discovery] of the number of days in a year. We will know the things about which the ancients doubted and thought differently, and we will know it from what they narrated, especially Hipparchus, a man of good knowledge, and *lover of knowledge* and truth.[13]

These "praises" are mainly found in the third and ninth books of the *Almagest*. However, they provide only weak evidence that Galileo had direct knowledge of a significant portion of the *Almagest* through

---

[10] The *Suda* is a known reference among historians of sixteenth-century science. Dilwyn Knox, for example, highlighted its importance as one of Copernicus' sources. Further details in Knox 2005, translated into Italian with a few additions in Knox 2013.

[11] "Ἵππαρχος, Νικαεύς, φιλόσοφος, γεγονὼς ἐπὶ τῶν ὑπάτων" (Adler 1928–1938, vol. II, p. 657, entry 521, emphasis added; English translation by Mary Pendergraft). Ada Adler noted that "ἐπὶ τῶν ὑπάτων" is corrupted (see ibid.). Special thanks to Flavio Bevacqua for bringing the *Suda* to my attention as a potential source for Galileo.

[12] See EN, VI, p. 340.

[13] "Primum autem omnium que oportet nos demonstrare de causa Solis est inventio scientie quantitatis longitudinis anni et numeri dierum eius. Ea vero in quibus dubitaverunt et diversificati sunt antiqui sciemus ex eis que ipsi narraverunt, et precipue Abrachis, vir bone scientie et *amator scientie* et veritatis" (Ptolemy 1515, ff. 26r-v, emphasis added).

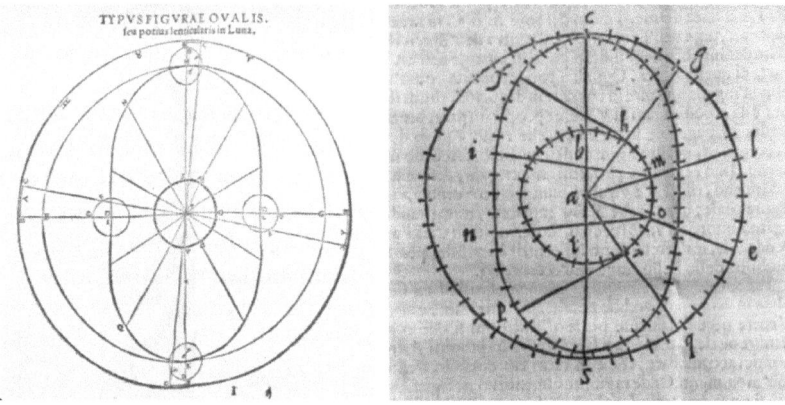

Fig. 5.1 Comparison of the *theoricae* of the Moon by Reinhold (on the left) and Mauri (on the right)

Gerard of Cremona's translation. It is also possible that Galileo referred to another source where Hipparchus is called a philosopher and remembered as a man praised by Ptolemy "throughout the *Almagest*."

In summary, the reference to "Hipparchus the philosopher" does not clarify how Galileo engaged with the more technical parts of the *Almagest*. Nonetheless, the study presented in this book may offer other valuable insights.

In Chapter 4, it was noted that Galileo likely contributed to some parts of Alimberto Mauri's *Considerations*. This work explicitly references Erasmus Reinhold's commentary on Peurbach's *Theoricae novae*, specifically his explanation of the *theorica Lunae*. According to Reinhold, the center of the Moon's epicycle follows a non-circular trajectory, namely an oval, or, better yet, a *lenticular* trajectory.[14] Interestingly, the diagram used by Reinhold to illustrate this non-circular motion differs from the one presented in Mauri's *Considerations* (see Fig. 5.1).[15]

If Galileo authored this exposition, it suggests that he studied the technical parts of the *Almagest* through Reinhold's commentary on Peurbach's *Theoricae novae*, a common approach at the time. We already

---

[14] Reinhold himself emphasizes the distinction between "oval" and "lenticular." The oval shape is specific to Mercury's *theorica*.

[15] See Mauri 1606, f. 13r.

know that Galileo taught at least one course in Padua on the *theoricae planetarum* and likely did so earlier in Pisa.[16]

Further exploration of these and other aspects is essential for developing a more comprehensive understanding of how Galileo acquired his knowledge of Ptolemaic astronomy.

[16] See *supra*, Chapter 2, n. 2; Chapter 3, n. 3.

Bibliography

Texts

Adler, Ada (ed.) (1928–1938). *Suidae Lexicon.* Leipzig: G. B. Teubner.

Al-Farghani, & al-Battani (1537). *Rudimenta Alfragrani. Item Albategnius astronomus peritissimus De motu stellarum, ex observationibus tum proprijs, tum Ptolemaei, omnia cum demonstrationibus Geometricis et Additionibus Ioannis de Regiomonte.* [...] *Norimberge.*

Alhacen, & Witelo (1572). *Opticae Thesaurus. Alhazeni Arabis libri septem, nunc primum editi. Eiusdem liber De crepusculis et Nubium ascensionibus. Item Vitellonis Thuringopoloni libri X. Omnes instaurati, figuris illustrati et aucti, adiectis etiam in Alhazenum commentarijs. A Federico Risnero.* [...] *Basileae, Per Episcopios.*

Amico, Giovan Battista (1536), *Ioannis Baptistae Amici Cosentini de Motibus corporum coelestium iuxta principia peripatetica sine eccentricis et epicyclis.* [...] *Impressi Venetiis a Ioanne Patavino et Venturino Roffinello.*

Barozzi, Francesco (1585). *Cosmographia In Quatuor Libros distributa* [...]. *Venetijs, ex Officina Gratiosi Perchacini.*

Barozzi, Francesco (1607). *Cosmografia in Quattro Libri divisa* [...]. *In Venetia, Presso Gratioso Perchacino.*

Biancani, Giuseppe (1620). *Sphaera mundi seu Cosmographia Demonstrativa, ac facili Methodo tradita.* [...] *Bononiae, Typis Sebastiani Bonomij. Superiorum permissu. Sumptibus Hieronymi Tamburini.*

Castelli, Benedetto (1669). *Alcuni opuscoli filosofici del padre abbate D. Benedetto Castelli da Brescia... In Bologna, per Giacomo Monti.*

I. Malara, *Galileo and the* Almagest, *c.1589–1592*, Palgrave Studies in the History of Science and Technology, https://doi.org/10.1007/978-3-031-70614-1

Castelli, Benedetto (2018). *Della misura dell'acque correnti; Alcuni opuscoli filosofici* (2nd ed.). Ed. by S. Ricciardo. Brescia: Morcelliana.

Clavius, Christoph (1570). *In Sphaeram Ioannis de Sacro Bosco commentarivs.* [...] *Romae,* [...] *Apud Victorium Helianum.*

Clavius, Christoph (1581), *In Sphaeram Ioannis de Sacro Bosco commentarius nunc iterum ab ipso auctore recognitus, et multis ac varijs locis locupletatus* [...]. *Romae* [...] *Ex Officina Dominici Basae.*

Clavius, Christoph (1585). *In Sphaeram Ioannis de Sacro Bosco commentarius nunc tertio ab ipso Auctore recognitus, et plerisque in locis locupletatus* [...] *Romae* [...] *Ex Officina Dominici Basae.*

Clavius, Christoph (1593). *In Sphaeram Ioannis de Sacro Bosco commentarius nunc quarto ab ipso Auctore recognitus,et plerisque in locis locupletatus. Lugduni. Sumptibus fratrum de Gabiano.*

Clavius, Christoph (1602). *In Sphaeram Ioannis de Sacro Bosco commentarius nunc quarto ab ipso Auctore recognitus,et plerisque in locis locupletatus. Lugduni. Sumptibus fratrum de Gabiano.1606: Christophori Clavii... Geometria practica.* [...] *Moguntiae, ex Typographo Ioannis Albini.*

Clavius, Christoph (1607). *In Sphaeram Ioannis de Sacro Bosco commentarius nunc quinto ab ipso Auctore hoc anno 1606 recognitus, et plerisque in locis locupletatus.* [...] *Romae. Sumptibus Io. Pauli Gellii ad signum navis.* [...] *Apud Aloisium Zanettum.*

Clavius, Christoph (1608). *In Sphaeram Ioannis de Sacro Bosco commentarius nunc postremo ab ipso Auctore recognitus, et plerisque in locis locupletatus.* [...] *S. Gervasii. Apud Samuelem Crispinum.*

Clavius, Christoph (1611). *Operum mathematicorum Tomus Tertius Complectens Commentarium in Sphaeram Ioannis di Sacro Bosco* [...] *Moguntiae, Sumptibus Antonii Hierat excudebat Reinhardus Eltz.*

Clavius, Christoph (1992). *Corrispondenza.* Critical ed. by U. Baldini, and P. D. Napolitani, 7 vols. Pisa: Università di Pisa.

Copernicus, Nicolaus (2015). *De revolutionibus orbium coelestium; Des révolutions des orbes célestes.* Critical edition, translation, and notes by M.-P. Lerner, A.-Ph. Segonds, and J.-P. Verdet, 3 vols. Paris: Les Belles Lettres.

Danti, Piervincenzo (1571). *La Sfera di Messer Giovanni Sacrobosco tradotta emendata et distinta in Capitoli da Piervincentio Dante de Rinaldi con molte et utili Annotationi del Medesimo. Rivista da Frate Egnatio Danti Cosmografo del Gran Duca di Toscana.* [...] *In Fiorenza, Nella Stamperia de Giunti.*

Della Porta, Giambattista (2000). *Claudii Ptolemaei Magnae constructionis liber primus, cum Theonis Alexandrini commentariis Io. Baptista Porta Neapolitano interprete.* Ed. by R. De Vivo, vol. XVIII of *Edizione nazionale delle opere di Giovan Battista della Porta.* Napoli: Edizioni scientifiche italiane.

Fracastoro, Girolamo (1538). *Hieronymi Fracastorii Homocentrica. Eiusdem De causis criticorum dierum per ea quae in nobis sunt. Venetiis.*

Frisius, Gemma (1545). *De Radio Astronomico et Geometrico liber* [...]. *Antuerpiae apud Greg. Bontium sub scuto Basilensi, et Lovanii Apud petrum Phalesium.*

Galilei, Galileo (1718). *Opere di Galileo Galilei Nobile Fiorentino Primario Filosofo, e Matematico del Serenissimo Gran Duca di Toscana.* 3 vols. Firenze: per Gio. Gaetano Tartini e Santi Franchi.

Galilei, Galileo (1744). *Opere di Galileo Galilei divise in quattro tomi, in questa nuova Edizione accresciute di molte cose inedite.* Padova: nella Stamperia del Seminario appresso Gio. Manfrè.

Galilei, Galileo (1960). *On Motion and on Mechanics: Comprising* De Motu *(ca. 1590) Translated with Introduction and Notes by I. E. Drabkin, and* Le Meccaniche *(ca. 1600) Translated with Introduction and Notes by Stillman Drake.* Madison: The University of Wisconsin Press.

Galilei, Galileo (1969). *Mechanics in Sixteenth-Century Italy: Selections from Tartaglia, Benedetti, Guido Ubaldo, and Galileo. Translated and Annotated by Stillman Drake and I. E. Drabkin.* Madison: The University of Wisconsin Press.

Galilei, Galileo (1965). *Il Saggiatore.* Ed. by L. Sosio. Milano: Feltrinelli.

Galilei, Galileo (1976). *Galileo against the philosophers in his* Dialogue of Cecco Ronchitti *(1605) and* Considerations of Alimberto Mauri *(1606).* English translations with introductions and notes by S. Drake. Los Angeles: Zeitlin & Ver Brugge.

Galilei, Galileo (1992). *Sidereus nuncius; Le messager celeste.* Text, translation, and notes by I. Pantin. Paris: Les Belles Lettres.

Galilei, Galileo (1998). *Dialogo sopra i due massimi sistemi del mondo tolemaico e copernicano.* Critical ed. and commentary by O. Besomi and M. O. Helbing, 2 vols. Padova: Antenore.

Galilei, Galileo (2009). *Scienza e religione. Scritti copernicani.* Ed. by M. Bucciantini and M. Camerota. Roma: Donzelli editore.

Galilei, Galileo (2023). *Il Saggiatore.* Ed. by M. Camerota and F. Giudice. Milano: Hoepli.

Geber (1534). *Libri IX de Astronomia* [...]. *Norimbergae apud Petreium.*

Mabillon, Jean (1687). *Iter Italicum litterarium... annis MDCLXXXV et MDCLXXXVI. Luteciae Parisiorum. Apud Viduam Edmund*[i] *Martin, Johannem Boudot et Stephanum Martin, in via Jacobea ad Solem aureum.*

Mauri, Alimberto (1606). *Considerazione d'Alimberto Mauri sopra alcuni luoghi del Discorso di Lodovico delle Colombe intorno alla stella apparita 1604. In Firenze. Appresso Gio. Antonio Caneo.*

Mazzoni, Jacopo (1597). *In Universam Platonis, et Aristotelis Philosophiam Praeludia, sive de Comparatione Platonis, et Aristotelis* [...]. *Venetiis Apud Ioannem Guerilium.*

Mazzoni, Jacopo (2010). *In universam Platonis et Aristotelis philosophiam Praeludia, sive de Comparatione Platonis et Aristotelis*. Ed. by S. Matteoli. Napoli: D'Auria.

Pappus of Alexandria (1588). *Pappi Alexandrini Mathematicae Collectiones a Federico Commandino Urbinate in Latinum Conversae, et commentariis illustrate. Pisauri, Apud Hieronymum Concordiam.*

Peurbach, Georg, & Regiomontanus (1496). *Epitoma Ioannis de monte regio in almagestum ptolemaei. [...] Explicit Magne Compositionis Astronomicon Epitoma Johannis de Regio monte. Impensis non nimis, curaque et emendatione non mediocri virorum prestantium Casparis Grossch, et Stephani Roemer. Opera quoque et arte impressionis mirifica viri solerti Johannis Hamman de Landoia, dictur Hertzog, felicibus astris expletum. [...] In hemispherio Veneto [...].*

Peurbach, Georg, & Regiomontanus (1543). *Epitome, in Cl. Ptolemaei Magnam compositionem [...]. Basileae apud Henrichum Petrum.*

Piccolomini, Alessandro (1540). *De la Sfera del Mondo libri quattro in lingua toscana [...]. In Venetia al segno del Pozzo.*

Piccolomini, Alessandro (1561). *De la Sfera del Mondo di M. Alessandro Piccolomini Libri quattro, Nuovamente da lui emendati, et di molte aggiunte in diversi luoghi largamente ampliati [...]. In Venetia, per Giovanni Varisco, et compagni.*

Ptolemy, Claudius (1515). *Almagestum Cl. Ptolemei Pheludiensis Alexandrini astronomorum principis: opus ingens ac nobile omnes celorum motus continens. Felicibus astris eat in lucem: Ductu Petri Lichtenstein Coloniensis Germani [...]. Venetiis ex officina eiusdem litteraria.*

Ptolemy, Claudius (1528). *Almagestum seu Magnae Contructionis Mathematicae Opus plane divinum Latina donatum lingua ab Georgio Trapezuntio usquequaque doctissimo. Per Lucam Gauricum Neapolitanum divinae matheseos professorem egregium in alma urbe Veneta orbis regina recognitum [...]. Venetiis [...] Luceantonii Iunta officina.*

Ptolemy, Claudius (1549). *Mathematicae constructionis Liber primus graecae et latinae editus. Additae explicationes aliquot locorum ab Erasmo Rheinolt Salvendensi. Wittebergae ex Officina Iohannis Lufft.*

Ptolemy, Claudius (1551) *Claudii Ptolemaei Pelusiensis Alexandrini omnia quae extant opera, praeter Geographiam [...]. Basileae [...]. In officina Henrichi Petri.*

Ptolemy, Claudius, & Theon of Alexandria (1538). ΚΛ. ΠΤΟΛΕΜΑΙΟΥ Μεγάλης συντάξεως βιβλ. ΙΓ. ΘΕΩΝΟΣ ΑΛΕΞΑΝΔΡΕΩΣ [*sic*] εἰς τὰ αὐτὰ ὑπομνημάτων βιβλ. ΙΑ. *Claudii Ptolemaei Magnae Constructionis, Idest Perfectae coelestium motuum pertractationis, Lib. XIII. Theonis Alexandrini in eosdem Commentariorum Lib. XI. Basileae apud Ioannem Walderum.*

Ptolemy, Claudius, & Theon of Alexandria (1605). *Claudii Ptolemaei magnae constructionis liber primus. Cum Theonis Alexandrini commentariis. Io. Baptista Porta Neap. Interprete. Neapoli. Typis Foelicis Stelliolae, ad Portam Regalem.*

Riccioli, Giovanni Battista (1551). *Almagestum novum astronomiam veterem novamque complectens* [...]. *Bononiae, Ex Typographia Haeredis Victorij Benatij.*

Sarpi, Paolo (1996). *Pensieri naturali, metafisici e matematici.* Critical ed. and commentary by L. Cozzi and L. Sosio. Milano, Napoli: Riccardo Ricciardi Editore.

Targioni Tozzetti, Giovanni (ed.) (1780). *Atti e memorie inedite dell'Accademia del Cimento e notizie aneddote dei progressi delle scienze in Toscana: contenenti, secondo l'ordine delle materie e dei tempi, memorie, esperienze, osservazioni, scoperte e la rinnovazione della fisica celeste e terrestre cominciando da Galileo Galilei fino a Francesco Redi ed a Vincenzo Viviani inclusive. Pubblicate dal dottore Gio. Targioni Tozzetti.* In Firenze: si vende da Giuseppe Tofani stampatore e da Luigi Carlieri librajo.

Theon of Alexandria (1821). *Commentaire de Théon d'Alexandrie, sur le premier* [et second] *livre de la Composition mathématique de Ptolémée.* Ed. and translation by N. Halma, 2 vols. Paris: Merlin.

Theon of Alexandira (1931–1943). *Commentaires de Pappus et de Théon d'Alexandrie sur l'Almageste.* Critical ed. by A. Rome, 3 vols. Roma: Biblioteca vaticana.

STUDIES

Abattouy, Mohamed (2000). *Essais galiléens: recherches sur la genèse et le développement de la science de Galilée* (preprint). Berlin: Max Planck Institut für Wissenschaftsgeschichte.

Abdukhalimov, Bahrom (1999). Ahmad Al-Farghānī and his *Compendium of Astronomy. Journal of Islamic Studies*, 10(2), pp. 142–158.

Aiton, Eric John (1987). Peurbach's *Theoricae novae planetarum*: A Translation with Commentary. *Osiris*, Second Series, vol. III, pp. 5–44.

Ariotti, Piero E. (1973). A little known early seventeenth century treatise on vision: Benedetto Castelli's *Discorso sopra la Vista* (1639, 1669). *Annals of Science*, 30(1), pp. 1–30.

Baldini, Ugo (1991). La teoria astronomica in Italia durante gli anni della formazione di Galileo: 1560–1610. In *Alle origini della rivoluzione scientifica.* Ed. by P. Casini, pp. 39–67. Roma: Istituto della Enciclopedia Italiana fondata da Giovanni Treccani.

Baldini, Ugo (1992). Legem impone subactis. *Studi su filosofia e scienza dei gesuiti in Italia (1540–1632).* Roma: Bulzoni.

Baldini, Ugo (2003). The Academy of Mathematics of the Collegio Romano from 1553 to 1612. In *Jesuit Science and the Republic of Letters*. Ed. by M. Feingold, pp. 47–98. Cambridge (Massachusetts), London: The MIT Press.

Barreca, Francesco (2018). Luis de Granada's *Introduttione del simbolo della fede* as a Possible Source for Galileo in the Preparation of His Letter to Piero Dini of March 23, 1615. *Galilaeana*, 15, pp. 115–135.

Battistini, Andrea (2005). «Girandole» verbali e «severità di geometriche dimostrazioni». Battaglie linguistiche nel *Saggiatore*. *Galilaeana*, 2, pp. 87–106.

Bellé, Riccardo (2004). *Il De refractione di G. B. Della Porta: edizione critica*. Ph.D. thesis: Università degli Studi di Firenze.

Bellone, Enrico (2003). *La stella nuova. L'evoluzione e il caso Galilei*. Torino: Einaudi.

Bernard, Alain (2010). The Alexandrian school. Theon of Alexandria and Hypatia. In *The Cambridge History of Philosophy in Late Antiquity*. Ed. by Lloyd P. Gerson, vol. I, pp. 417–436. Cambridge: Cambridge University Press.

Bernard, Alain (2014). In What Sense Did Theon's Commentary on the *Almagest* Have a Didactic Purpose? In *Scientific Sources and Teaching Contexts Throughout History: Problems and Perspectives*. Ed. by A. Bernard and C. Proust, pp. 95–121. Dordrecht: Springer.

Biagioli, Mario (2003). Stress in the Book of Nature: The Supplemental Logic of Galileo's Realism. *MLN*, 118(3), pp. 557–585.

Biagioli, Mario (2006). *Galileo's Instruments of Credit: Telescopes, Images, Secrecy*. Chicago: The University of Chicago Press.

Bianchi, Luca (2022). Galileo e le «abilità diverse degl'intelletti»: note sulla dedica al *Dialogo sopra i due massimi sistemi*. In *Atti e memorie dell'Accademia toscana di scienze e lettere La Colombaria, vol. LXXXVI (N.S. - LXXII). Rivoluzione scientifica e tradizioni filosofiche per Maurizio Torrini. Atti del Seminario tenuto in Colombaria il 7 maggio 2021*. Ed. by L. Fonnesu and A. Savorelli, pp. 297–312. Firenze: Olschki.

Firenze 2022, pp. 297–312.

Blackwell, Richard J. (1991). *Galileo, Bellarmine, and the Bible*. Notre Dame: University of Notre Dame Press.

Bowen, Alan C., & Todd, Robert B. (eds.) (2004). *Cleomedes' lectures on astronomy: a translation of* The Heavens *with an introduction and commentary by Alan C. Bowen and Robert B. Todd*. Berkeley: University of California Press.

Bucciantini, Massimo (1992). Atomi geometria e teologia nella filosofia galileiana di Benedetto Castelli. In *Geometria e atomismo nella scuola galileiana*. Ed. by M. Bucciantini and M. Torrini, pp. 171–191. Firenze: Olschki.

Bucciantini, Massimo (1995). *Contro Galileo. All'origine dell'*Affaire. Firenze: Olschki.

Bucciantini, Massimo (2003). *Galileo e Keplero. Filosofia, cosmologia e teologia nell'Età della Controriforma.* Torino: Einaudi.

Bucciantini, Massimo, & Camerota, Michele, & Giudice, Franco (2015). *Galileo's Telescope: A European History.* Tr. by Bolton Catherine. Cambridge, MA: Harvard University Press. [Original: *Il telescopio di Galileo: una storia europea.* Torino: Einaudi, 2012]

Burnett, Charles (2010). ›Ptolemaeus in Almagesto dixit‹: The Transformation of Ptolemy's *Almagest* in its Transmission via Arabic into Latin. In *Transformationen antiker Wissenschaften.* Ed. by G. Toepfer and H. Böhme, pp. 115–140. Berlin, New York: De Gruyter.

Burnett, Charles (2013). Translation and Transmission of Greek and Islamic Science to Latin Christendom. In *The Cambridge History of Science.* Ed. by D.C. Lindberg and M.H. Shank, vol. II, pp. 341–364. Cambridge: Cambridge University Press.

Büttner, Jochen (2008). Big Wheel Keep on Turning. *Galilaeana*, 5, pp. 33–62.

Büttner, Jochen (2019). *Swinging and Rolling. Unveiling Galileo's Unorthodox Path from a Challenging Problem to a New Science.* Dordrecht: Springer.

Camerota, Michele (1991). Movimento circolare e *motus neuter* negli scritti negli scritti *De motu* di Galileo Galilei. *Annali della Facoltà di Magistero dell'Università di Cagliari*, 15(1), pp. 185–221.

Camerota, Michele (1992). *Gli scritti* De motu antiquiora *di Galileo Galilei. Il Ms. Gal. 71: un'analisi storico-critica.* Cagliari: CUEC.

Camerota, Michele (2000). I *De motu antiquiora* di Galileo e il *Discorso idrostatico* del 1612: affinità e differenze. *Annali della Facoltà di scienze della formazione dell'Università di Cagliari*, 23, pp. 75–98.

Camerota, Michele (2004). *Galileo Galilei e la cultura scientifica nell'Età della Controriforma.* Rome: Salerno Editrice.

Camerota, Michele (2016). Buonamici, Francesco. In *Encyclopedia of Renaissance Philosophy.* Ed. by M. Sgarbi. Cham: Springer.

Camerota, Michele, & Helbing, Mario Otto (2000). Galileo and the Pisan Aristotelianism: Galileo's «De motu antiquiora» and the *Quaestiones de Motu Elementorum* of the Pisan Professors. *Early Science and Medicine*, 5(4), pp. 141–175.

Cardoso, Walmir Thomazi, & De Andrade Martins, Roberto (2008). O *Trattato della Sfera ovvero Cosmografia* de Galileo Galilei e algumas Cosmografias e Tratados da Esfera do Século XVI. *Episteme*, 13(27), pp. 15–38.

Cardoso, Walmir Thomazi, & De Andrade Martins, Roberto (2017). Galileo's *Trattato della sfera ovvero cosmografia* and Its Sources. *Philosophia Scientiae*, 21(1), pp. 131–147.

Carolino, Luís Miguel (2023). How did a Lutheran astronomer get converted into a Catholic authority? The Jesuits and their reception of Tycho Brahe in Portugal. *The British Journal for the History of Science*, pp. 1–22.

Carugo, Adriano, & Crombie, Alistair C. (1983). The Jesuits and Galileo's Ideas of Science and of Nature. *Annali dell'Istituto e Museo di storia della scienza di Firenze*, 8(2), pp. 3–67.

Caverni, Raffaello (1891–1900). *Storia del metodo sperimentale in Italia*, 6 vols. Firenze: Stab. G. Civelli.

Clavelin, Maurice (1996). *La philosophie naturelle de Galilée* (2nd ed.). Paris: Albin Michel.

Clavelin, Maurice (2016). *Galilée, cosmologie et science du mouvement suivi de Regards sur l'empirisme au XXe siècle*. Paris: CNRS Editions.

Dear, Peter R. (2001). *Revolutionizing the sciences: European knowledge and its ambitions, 1500–1700*. Basingstoke: Palgrave.

De Andrade Martins, Roberto (2010). Galileo Galilei, los climas y la tradición del *Tractatus de sphaera* de Johannes de Sacrobosco. In *Epistemología e Historia de la Ciencia. Selección de Trabajos de las XX Jornadas. Facultad de Filosofía y Humanidades*. Ed. by P. García and A. Massolo, pp. 373–380. Córdoba: Universidad Nacional de Córdoba, Argentina.

De Pace, Anna (1993). *Le matematiche e il mondo. Ricerche su un dibattito in Italia nella seconda metà del Cinquecento*. Milano: FrancoAngeli.

De Pace, Anna (2005). Galileo interprete del Timeo. In *Storia della scienza, storia della filosofia: Interferenze*. Ed. by G. Canziani, pp. 39–76. Milano: FrancoAngeli.

De Pace, Anna (2009). *Niccolò Copernico e la fondazione del cosmo eliocentrico. Con testo, traduzione e commentario del Libro I de* Le rivoluzioni celesti. Milano: Mondadori.

De Pace, Anna (2020). *Galileo lettore di Copernico*. Firenze: Olschki.

Di Bono, Mario (1990). *Le sfere omocentriche di Giovan Battista Amico nell'astronomia del Cinquecento: con il testo del* «De motibus corporum coelestium...». Genova: C.N.R., Centro di studio sulla storia della tecnica.

Di Bono, Mario (1995). Copernicus, Amico, Fracastoro and Tusi's device. *Journal for the history of astronomy*, 26, pp. 133–154.

Di Bono, Mario (2006). Temi e fonti sull'*Homocentrica* di Fracastoro. In *Girolamo Fracastoro: fra medicina, filosofia e scienze della natura. Atti del Convegno internazionale di studi in occasione del 450° anniversario della morte, Verona-Padova, 9–11 ottobre 2003*. Ed. by A. Pastore and E. Peruzzi, pp. 117–140. Firenze: Olschki.

Dollo, Corrado (2003). *Gaileo Galilei e la cultura della tradizione*. Ed. by G. Bentivegna, S. Burgio and G. Magnano San Lio. Soveria Mannelli: Rubettino.

Drabkin, Israel E. (1960). A note on Galileo's *De motu*. *Isis*, 51, pp. 270-277.

Drake, Stillman (1977). Tartaglia's *squadra* and Galileo's *compasso*. *Annali dell'Istituto e Museo di storia della scienza di Firenze*, 2(1), pp. 35-54.

Drake, Stillman (1978). *Galileo at Work*. Chicago: Chicago University Press.

Drake, Stillman (1987). Galileo's Steps to Full Copernicanism, and Back. *Studies in History and Philosophy of Science*, 18(1), pp. 93–105.

Drake, Stillman (1999). Ptolemy, Galileo, and Scientific Method. In Id., *Essays on Galileo and the History and Philosophy of Science*, vol. I, pp. 273–292. Toronto: University of Toronto Press. [Repr. from *Studies in History and Philosophy of Science*, 9(2), pp. 99–115]

Dupré, Sven (2002). *Galileo, the Telescope and the Science of Optics in the Sixteenth Century: A Case Study of Instrumental Practice in Art and Science*. Ph.D. thesis. Universiteit Gent.

Ekler, Péter (2019). Georgius Trapezuntius, Johannes Regiomontanus and the *Defensio Theonis*. In *Bysanz und das Abendland VI. Studia byzantino-occidentalia*. Ed. by Z. Farkas, L. Horvát, T. Mészáros, pp. 211–218. Budapest: Eötvös-József-Collegium.

Enenkel, Karl A.E. (2014). Introduction. The Transformation of the Classics. Practices, Forms, and Functions of Early Modern Commenting. In *Intersections: Interdisciplinary Studies in Early Modern Culture*. Ed. by K.A.E. Enenkel, pp. 1–12. Leiden, Boston: Brill.

Enenkel, Karl A.E., & Nellen, Henk (2013). Introduction. Neo-Latin Commentaries and the Management of Knowledge. In *Neo-Latin Commentaries and the Management of Knowledge in the Late Middle Ages and the Early Modern Period (1400 -1700)*. Ed. by K.A.E. Enenkel and H. Nellen, pp. 1–76. Leuven: Leuven University Press.

Evans, James (1998). *The History and Practice of Ancient Astronomy*. New York, Oxford: Oxford University Press.

Evans, James (2018). Ptolemy. In *The Oxford Handbook of Science and Medicine in the Classical World*. Ed. by P.T. Keyser and J. Scarborough, pp. 789–927. New York: Oxford University Press.

Favaro, Antonio (1966). *Galileo Galilei e lo Studio di Padova*, 2 vols. Padova: Antenore.

Favaro, Antonio (1886). *La libreria di Galileo Galilei*. Bologna: Forni.

Feke, Jacqueline (2018). *Ptolemy's Philosophy: Mathematics as a Way of Life*. Princeton: Princeton University Press.

Festa, Egidio, & Roux, Sophie (2006). La moindre petite force peut mouvoir un corps sur un plan horizontal: l'emergence d'un principe mécanique et son devenir cosmologique. *Galilaeana*, 3, pp. 123–147.

Finocchiaro, Maurice A. (1989). *The Galileo Affair: A Documentary History*. Berkeley: University of California Press.

Finocchiaro, Maurice A. (2005). *Retrying Galileo, 1633–1992*. Berkeley, Los Angeles, London: University of California Press.

Forcada, Miquel (2014). Ibn Rushd: Abū al-Walīd Muḥammad Ibn Aḥmad Ibn Muḥammad Ibn Rushd al-Ḥafīd. In *Biographical Encyclopedia of Astronomers*. Ed. by T. Hockey et al. New York (NY): Springer.

Fredette, Raymond (1969). *Les* De motu *plus anciens de Galileo Galilei: prolegomenes*. Ph.D. thesis. Université de Montreal.

Fredette, Raymond (1972). Galileo's *De motu antiquiora*. *Physis*, 14(4), pp. 321–348.

Fredette, Raymond (2001). Galileo's *De Motu Antiquiora*. Notes for a Reappraisal. In *Largo campo di filosofare. Eurosymposium Galileo 2001*. Ed. by J. Montesino and C. Solis, pp. 165–181. La Orotava: Fundacion Canaria Orotava de Historia de la Ciencia.

Fredette, Raymond (2017). Les *De motu antiquiora* de Galileo Galilei: le lancement de la carrière du filosofo-geometra. *Philosophia scientiae*, 21(1), pp. 55–70.

Fuiano, Michele (1967). Astrologia ed umanesimo in due prefazioni di Andrea di Trebisonda. In *Estratto dagli "Atti" dell'Accademia Pontaniana, Nuova Serie, Volume XVII*. Napoli: Giannini.

Galluzzi, Paolo (1973). Il platonismo del tardo Cinquecento e la filosofia di Galileo. In *Ricerche sulla cultura dell'Italia moderna*. Ed. by P. Zambelli and M. Bellucci, pp. 37–79. Roma: Laterza.

Galluzzi, Paolo (1979). *Momento: studi galileiani*. Roma: Edizioni dell'Ateneo e Bizzarri.

Galluzzi, Paolo (2011). *Tra atomi e indivisibili: la materia ambigua di Galileo*. Firenze: Olschki.

Gamba, Enrico, & Montebelli, Vico (1988). *Le scienze a Urbino nel tardo Rinascimento*. Urbino: QuattroVenti

Giudice, Franco (2014). Galileo's Cosmological View from the *Sidereus Nuncius* to *Letters on Sunspots*. *Galilaeana*, 11, pp. 49–63.

Giusti, Enrico (1998). Elements for the relative chronology of Galilei's *De motu antiquiora*. *Nuncius*, 13(2), pp. 427–460.

Goldstein, Bernard R. (1977). Ibn Mu'adh's Treatise on Twilight and the Height of the Atmosphere. *Archive for History of Exact Sciences*, 17(2), pp. 97–118.

Goldstein, Bernard R. (1987). Remarks on Gemma Frisius's *De Radio Astronomico et Geometrico*. In *From Ancient Omens to Statistical Mechanics*, Ed. by J.L. Berggren and B.R. Goldstein, pp. 167–180. Copenhagen: University library.

Goldstein, Bernard R., & Smith, A. Mark (1993). The Medieval Hebrew and Italian Versions of Ibn Mu'adh's 'On Twilight and the Rising of Clouds.' *Nuncius*, 8(2), pp. 611–643.

Grafton, Anthony (1988). The Availability of Ancient Works. In *The Cambridge History of Renaissance Philosophy*, Ed. by C.B. Schmitt, and Q. Skinner, pp. 767–791. Cambridge: Cambridge University Press.

Grafton, Anthony (2010). Commentary. In *The Classical Tradition*. Ed. by A. Grafton, G.W. Most, and S. Settis, pp. 225–233. Cambridge, MA: Harvard University Press.

Granada, Miguel Á. (2023). Jerónimo Muñoz y Juan Cedillo Díaz: el Sol como "corazón del mundo" en el debate en torno a Copérnico. *Galilaeana*, 20(1), pp. 85–120.

Grant, Edward (1984). In Defense of the Earth's Centrality and Immobility: Scholastic Reaction to Copernicanism in the Seventeenth Century. *Transactions of the American Philosophical Society*, 74(4), pp. 1–69.

Grant, Edward (2003) The Partial Transformation of Medieval Cosmology by Jesuits in the Sixteenth and Seventeenth Centuries. In *Jesuit Science and the Republic of Letters*. Ed. by by M. Feingold, pp. 127–155. Cambridge (Massachusetts), London: The MIT Press.

Guerrini, Luigi (2011). *La scienza infedele*. Roma: Vecchiarelli.

Guerrini, Luigi (2014). Pereira and Galileo: Acceleration in Free Fall and Impetus Theory. *Bruniana et Campanelliana*, 20(2), pp. 513–530.

Hartner, Willy (1967). Galileo's Contribution to Astronomy. In *Galileo Man of Science*. Ed. by E. McMullin, pp. 178–194. New York, London: Basic Books.

Hatfield, Gary (1990). Metaphysics and the New Science. In *Reappraisals of the Scientific Revolution*. Ed. by D.C. Lindberg and R. Westman, pp. 93–166. Cambridge: Cambridge University Press.

Heilbron, John L. (2010). *Galileo*. Oxford: Oxford University Press.

Helbing, Mario Otto (1989). *La filosofia di Francesco Buonamici, professore di Galileo a Pisa*. Pisa: Nistri-Lischi.

Jones, Alexander (1999). Uses and Users of Astronomical Commentaries in Antiquities. In *Commentaries: Kommentare*. Ed. by G.W. Most, pp. 147–172. Göttingen: Vandenhoeck und Ruprecht.

Knox, Dilwyn (2005). Copernicus's Doctrine of Gravity and the Natural Circular Motion of the Elements. *Journal of the Warburg and Courtauld Institutes*, 68, pp. 157–211.

Knox, Dilwyn (2013). *Copernico e la dottrina della gravità. La dottrina della gravità e del moto circolare degli elementi nel* De revolutionibus. Pisa, Roma: Fabrizio Serra.

Koyré, Alexandre (1966). *Etudes Galiléennes*. Paris: Hermann. [First published in 1939]

Kren, Claudia (1968). Homocentric Astronomy in the Latin West. The *De reprobatione ecentricorum et epiciclorum* of Henry of Hesse. *Isis*, 59(3), pp. 269–281.

Kunitzsch, Paul (1974). *Der* Almagest: *Die* Syntaxis Mathematica *des Claudius Ptolemäus in arabisch-lateinischer Überlieferung*. Wiesbaden: Harrassowitz.

Kunitzsch, Paul (2008). *Almagest*: Its Reception and Transmission in the Islamic World. In *Encycloplaedia of the History of Science, Technology, and Medicine in non-Western Cultures*. Ed. by H. Selin, vol. I, pp. 140–141. Berlin: Springer.

Langermann, Y. Tzvi (2020). Revamping Ptolemy's Proofs for the Sphericity of the Heavens: Three Arabic Commentaries on *Almagest* I.3. In *Ptolemy's Science of the Stars in the Middles Ages*. Ed. by D. Juste, B. van Dalen, D. N. Hasse, and C. Burnett, pp. 159–180. Turnhout: Brepols.

Lattis, James M. (1994). *Between Copernicus and Galileo. Christoph Clavius and the Collapse of Ptolemaic Cosmology*. Chicago and London: The University of Chicago Press.

Lehn, Waldemar H., & van der Werf, Siebren (2005). Atmospheric Refraction: A History. *Applied Optics*, 44(27), pp. 5624–5636.

Lerner, Michel-Pierre (1995). L'entrée de Tycho Brahe chez les jésuites ou le chant du cygne de Clavius. In *Les jésuites à la Renaissance. Système éducatif et production du savoir*. Ed. by L. Giard, pp. 145–185. Paris: Press Universitaires de France.

Lerner, Michel-Pierre (1996). *Le Monde des sphères: Genèse et triomphe d'un représentation cosmique*, vol. I. Paris: Les Belles Lettres.

Lindberg, David C. (1983). *Studies in the History of Medieval Optics*. London: Variorum reprints.

Little, John B. (2022). The Eclectic Content and Sources of Clavius's *Geometria Practica*. *Archive for History of Exact Sciences*, 76(4), pp. 391–424.

Malara, Ivan (2019). Galileo and His Sources? A different Methodological Approach to Galileo's *Juvenilia*. *Galilaeana*, 16, pp. 1–40.

Malara, Ivan (2020) Gravitas, Renaissance Concept of. In *Encyclopedia of Renaissance Philosophy*. Ed. by M. Sgarbi. Cham: Springer.

Malara, Ivan (2021). *Galileo Creation and Cosmogoy: A Study on the Interplay between Galileo's Science of Motion and the Creation Theme*. Ph.D. thesis. Università degli Studi di Milano, Universiteit Gent.

Malara, Ivan (2023). Galileo as a Reader of Ptolemy: Notes on the Occasion of the 400th Anniversary of *Il Saggiatore* (1623). *Rivista di Storia della Filosofia*, 78(3), pp. 465–473.

Malpangotto, Michela (2008). *Regiomontano e il rinnovamento del sapere matematico e astronomico nel Quattrocento*. Bari: Cacucci.

Malpangotto, Michela (2021). *Theoricae novae planetarum Georgii Peurbachii dans l'histoire de l'astronomie*. Paris: CNRS.

Monfasani, John (1976). *George of Trebizond. A Biography and Study of his Rhetoric and Logic*. Leiden: Brill.

Monfasani, John (1984). *Collectanea Trapezuntiana: Texts, Documents, and Bibliographies of George of Trebizond*. Binghamton: Center for Medieval & Early Renaissance studies.

Most, Glenn W. (ed.) (1999). *Commentaries: Kommentare*. Göttingen: Vanden-hoeck und Ruprecht.

Navarro Brotons, Víctor (2019). *Jerónimo Muñoz: Matemáticas, cosmología y humanismo en la época del Renacimiento*. València: Universitat de València.

Neugebauer, Otto (1975). *A History of Ancient Mathematical Astronomy*, 3 vols. Berlin: Springer.

Neugebauer, Otto, & Swerdlow, Noel M. (1984). *The Mathematical Astronomy in Copernicus' De* Revolutionibus, *in Two Parts*. New York: Springer.

Omodeo, Pietro D. (2014). *Copernicus in the Cultural Debates of the Renaissance: Reception, Legacy, Transformation*. Leiden, Boston: Brill.

Omodeo, Pietro D., & Regier, Jonathan (2019). The Wittenberg Reception of Copernicus: At the Origin of a Scholarly Tradition. In *Natural knowledge and Aristotelianism at Early Modern Protestant Universities*. Ed. by P.D. Omodeo, V. Wels, pp. 83–108. Wiesbaden: Harrassowitz.

Omodeo, Pietro D., & Tupikova, Irina (2013). The Post-Copernican Reception of Ptolemy: Erasmus Reinhold's Commented Edition of the *Almagest*, Book One (Wittenberg, 1549). *Journal for the History of Astronomy*, 44(3), pp. 235–256.

Omodeo, Pietro D., & Tupikova, Irina (2016). Cosmology and Epistemology: A Comparison between Aristotle's and Ptolemy's Approaches to Geocentrism. In *Spatial Thinking and External Representation: Towards a Historical Epistemology of Space*. Ed. by M. Schemmel, pp. 145–174. Edition Open Access, Max Planck Institute for the History of Science.

Omodeo, Pietro D., & Tupikova, Irina (2018). Visual and Verbal Commentaries in Renaissance Astronomy: Erasmus Reinhold's Treatment of Classical Sources on Astronomy. *Philological Encounters*, 3(3), pp. 359–398.

Pagano, Sergio (ed.) (1984). *I documenti del processo di Galileo Galilei*. Città del Vaticano: Archivio Vaticano.

Pagano, Sergio (ed.) (2009). *I documenti vaticani del processo di Galileo Galilei*. Città del Vaticano: Archivio segreto vaticano.

Pantin, Isabelle (2012). The First Phases of the *Theoricae Planetarum* Printed Tradition (1474–1535): The Evolution of a Genre Observed Through Its Images. *Journal for the History of Astronomy*, 43(1), pp. 3–26.

Pedersen, Olaf (2011). *A Survey of the Almagest. With Annotation and New Commentary by Alexander Jones*. New York (NY): Springer.

Pellacani, Daniele (2020). Le edizioni dell'«Almagesto» nel XVI secolo, e un esemplare postillato da Ercole Bottrigari. *Ecdotica*, 17, pp. 37–74.

Piccinali, Luca (2018). *La formazione scientifica di Benedetto Castelli tra Brescia e Padova (1578–1610)*. MA thesis. Università degli Studi di Trento, Università degli Studi di Verona.

Piccolino, Marco, & Wade, Nicholas J. (2013). *Galileo's Visions: Piercing the Spheres of the Heavens by Eye and Mind*. Oxford: Oxford University Press.

Plug, Cornelis, & Ross, Helen E. (1989). Historical Review. In *The Moon Illusion*. Ed. by M. Hershenson, pp. 5–28. Hillsdale: Lawrence Erlbaum Associates.

Purnell, Frederik Jr. (1971). *Jacopo Mazzoni and His Comparison of Plato and Aristotle*. Ph.D. thesis: Columbia University.

Purnell, Frederik Jr. (1972). Jacopo Mazzoni and Galileo. *Physis*, 14, pp. 273–294.

Reeves, Eileen (2008). *Galileo's Glassworks: The Telescope and the Mirror*. Cambridge, MA: Harvard University Press.

Renn, Jürgenn (1992). Proofs and Paradoxes: Free Fall and Projectile Motion in Galileo's physics. In *Exploring the Limits of Preclassical Mechanics. A Study of Conceptual Development in Early Modern Science: Free Fall and Compounded Motion in the Work of Descartes, Galileo, Beeckman*. Ed. by P. Damerow, G. Freudenthal, P. McLaughlin, J. Renn, pp. 126–276. New York: Springer.

Renn, Jürgen, et al. (2001). Hunting the White Elephant: Ehen and How Did Galileo DIscover the Law of Fall. In *Galileo in Context*. Ed. by J. Renn, pp. 29–149. Cambridge: Cambridge University Press.

Romana Berno, Francesca (2008). Appunti sul latino di Galileo Galilei. In *Atti e Memorie dell'Accademia Galileiana di Scienze, Lettere ed Arti già dei Ricovrati e Patavina*, v. 119(3), pp. 15–37. Padova: La Garangola.

Rose, Paul L. (1971). Plusieurs manuscrits autographes de Federico Commandino à la Bibliothèque Nationale de Paris. *Revue d'histoire des sciences*, 24(4), pp. 299–307.

Rosen, Edward (1992). *Nicholas Copernicus on the Revolutions*. Translation and commentary by E. Rosen. Baltimore, London: John Hopkins University Press.

Ross, Helen E. (2000). Cleomedes (c. 1st century AD) on the celestial illusion, atmospheric enlargement, and size-distance invariance. *Perception*, 29(7), pp. 862–871.

Ross, Helen E., & Ross, George M. (1976). Did Ptolemy understand the moon illusion? *Perception*, 5(4), pp. 377–385.

Sabra, Abdelhamid I. (1967). The Authorship of the *Liber de crepusculis*, an Eleventh-Century Work on Atmospheric Refraction. *Isis*, 58(1), pp. 77–85.

Schmitt, Charles B. (1972). The Faculty of Arts at Pisa at the Time of Galileo. *Physis*, 14, pp. 243–272. [Repr. in Id., *Studies in Renaissance Philosophy and Science*, chapter IX. London, 1981: Variorum Reprints]

Schmitt, Charles B. (1978). Filippo Fantoni, Galileo Galilei's Predecessor as Mathematics Lecturer at Pisa. In *Science and history: studies in honor of Edward Rosen*, pp. 53–62. Wroclaw: Ossolineum. [Repr. in Id., *Studies in Renaissance Philosophy and Science*, chapter X. London, 1981: Variorum reprints]

Shank, Michael H. (1998). Regiomontanus and Homocentric Astronomy. *Journal for the History of Astronomy*, 29(2), pp. 157–166.

Shank, Michael H. (2002). Regiomontanus on Ptolemy, Physical Orbs, and Astronomical Fictionalism. Goldsteinian Themes in the *Defense of Theon against George of Trebizond. Perspectives on Science*, 10(2), pp. 179–207.

Shank, Michael H. (2007). Regiomontanus as a Physical Astronomer. Samplings from the *Defence of Theon against George of Trebizond. Journal for the History of Astronomy*, 38(3), pp. 325–349.

Shank, Michael H. (2017). The *Almagest*, Politics, and Apocalypticism in the Conflict between George of Trebizond and Cardinal Bessarion. *Almagest*, 8(2), pp. 49–83.

Shank, Michael H. (2020). Regiomontanus *versus* George of Trebizond on Planetary Order, Distances, and Orbs (*Almagest* 9.1). In *Ptolemy's Science of the Stars in the Middle Ages*. Ed. by D. Juste, B. van Dalen, D. N. Hasse, and C. Burnett, pp. 305–386. Turnhout: Brepols.

Sisana, Beatrice (2023). *'L'archimedeismo' negli scritti giovanili di Galileo*. Ph.D. thesis. Università Roma Tre.

Smith, A. Mark (1992). The Latin Version of Ibn Mu'ahd's Treatise 'On Twilight and the Rising of Clouds.' *Arabic Sciences and Philosophy*, 2(1), pp. 38–88.

Smith, A. Mark (2003). Ptolemy, Alhacen, and Ibn Mu'adh and the Problem of Atmospheric Refraction. *Centaurus*, 45(1–4), pp. 100–115.

Smith, A. Mark (2010). *Alhacen on Refraction. A Critical Edition, with English Translation and Commentary, of Book 7 of Alhacen's* De Aspectibus, *the Medieval Latin Version of Ibn al-Haythan's* Kitāb al-Manāzir, 2 vols. *Transactions of the American Philosophical Society*, 100(3), sections 1 and 2.

Smith, A. Mark (2017). *Optical Magic in the Late Renaissance: Giambattista Della Porta's "De Refractione." Transactions of the American Philosophical Society*, New Series, 107(1).

Swerdlow, Noel M. (1972). Aristotelian Planetary Theory in the Renaissance: Giovanni Battista Amico's Homocentric Spheres. *Journal for the History of Astronomy*, 3(1), pp. 36–48.

Swerdlow, Noel M. (1999). Regiomontanus's Concentric-Sphere Models for the Sun and the Moon. *Journal for the History of Astronomy*, 30(1), pp. 1–23.

Swerdlow, Noel M. (2021). Unpublished manuscript.

Taub, Liba C. (1993). *Ptolemy's Universe: The Natural Philosophical and Ethical Foundations of Ptolemy's Astronomy*. Chicago: Open Court.

Thorndike, Lynn (1949). *The* Sphere *of Sacrobosco and Its Commentators*. Chicago: University of Chicago Press.

Toomer, Gerald J. (1976). Theon of Alexandria. In *Dictionary of Scientific Biography*, ed. by C. C. Gillispie, vol. XIII, pp. 312–325. New York (NY): Charles Scribner's Sons.

Toomer, Gerald J. (1984). *Ptolemy's* Almagest. New York (NY): Springer.

Valleriani, Matteo (2010). *Galileo Engineer*. Dordrecht: Springer.

Valleriani, Matteo (ed.) (2020). De sphaera *of Johannes de Sacrobosco in the Early Modern Period: The Authors of the Commentaries*. Cham: Springer.

Valleriani, Matteo, & Ottone, Andrea (eds.) (2022). *Publishing Sacrobosco's* De Sphaera *In Early Modern Europe: Modes of Material and Scientific Exchange*. Cham: Springer.

Van Dyck, Maarten (2006). *An Archeology of Galileo's Science of Motion*. Ph.D. thesis. Universiteit Gent.

Van Dyck, Maarten (2018). Idealization and Galileo's Proto-Inertial Principle. *Philosophy of Science*, 85(5), pp. 919–929.

Van Dyck, Maarten (2022a). Applying Mathematics to Nature. In *The Cambridge History of Philosophy of the Scientific Revolution*. Ed. by D. M. Miller and D. Jalobeanu, pp. 254–273. Cambridge: Cambridge University Press.

Van Dyck, Maarten (2022b). Galilean Challenges: An Essay Review of Jochen Büttner, *Swinging and Rolling. Unveiling Galileo's Unorthodox Path from a Challenging problem to a New Science. Centaurus*, 64(4), pp. 925–940.

Van Dyck, Maarten, & Malara, Ivan (2022). Impetus, Renaissance Concept of. In *Encyclopedia of Renaissance Philosophy*. Ed. by M. Sgarbi. Cham: Springer.

Verardi, Donato (2015). Della Porta, Giambattista. In *Encyclopedia of Renaissance Philosophy*. Ed. by M. Sgarbi. Cham: Springer.

Wallace, William A. (1977). *Galileo's Early Notebooks: The Physical Questions. A Translation from the Latin with Historical and Paleographical Commentary*. Notre Dame: University of Notre Dame Press.

Wallace, William A. (1981). *Prelude to Galileo. Essays on Medieval and Sixteenth-Century Sources of Galileo's Thought*. Dordrecht, Boston, London: Reidel.

Wallace, William A. (1984a). Galileo's Early Arguments for Geocentrism and His Later Rejection of Them. In *Novità celesti e crisi del sapere: Atti del Convegno internazionale di studi galileiani*. Ed. by P. Galluzzi, pp. 31–40. Firenze: Giunti Barbèra.

Wallace, William A. (1984b). *Galileo and His Sources: The Heritage of the Collegio Romano in Galileo's Science*. Princeton: Princeton University Press.

Wohlwill, Emil (1884). Über die Entdeckung des Beharrungsgesetzes. *Zeitschrift für Völkerpsychologie und Sprachwissenschaft*, 15, pp. 70–135, 337–387.

Wohlwill, Emil (1904). Melanchthon und Copernicus. *Mitteilungen zur Geschichte der Medizin und der Naturwissenschaft*, 3, pp. 260–267.

Wohlwill, Emil (1909). *Galilei und sein Kampf für die copernicanische Lehre*, 2 vols. Hamburg und Leipzig: L. Voss.

Wolff, Michael (1987). Impetus Mechanics as a Physical Argument for Copernicanism: Copernicus, Benedetti, Galileo. *Science in Context*, 1(2), pp. 215–256.

Zepeda, Henry (2015). Euclidization in the *Almagest Parvum. Early Science and Medicine*, 20(1), pp. 48–76.

Zepeda, Henry (2018). *The First Latin Treatise on Ptolemy's Astronomy: the Almagesti minor (c. 1200)*. Turnhout: Brepols.

Zinner, Ernst (1990). *Regiomontanus: His Life and Work*. Tr. E. Brown. Amsterdam, New York, Oxford, Tokyo: North-Holland. [Original: *Leben und Wirken des Joh. Müller von Königsberg, genannt Regiomontanus*. Osnabrück: O. Zeller, 1968.]

WEB SOURCES

Juste, David [3]. 'Ptolemy, *Almagesti* (tr. Gerard of Cremona)' (update: 30.07.2023), *Ptolemaeus Arabus et Latinus. Works*, URL = http://ptolem aeus.badw.de/work/3.

Juste, David [19]. 'Ptolemy, *Almagest* (Greek)' (update: 17.01.2024), *Ptolemaeus Arabus et Latinus. Works*, URL = http://ptolemaeus.badw.de/work/19.

Juste, David [70]. 'Geber, *Liber super Almagesti*' (update: 03.07.2023), *Ptolemaeus Arabus et Latinus. Works*, URL = http://ptolemaeus.badw.de/work/70.

Juste, David [83]. 'Theon of Alexandria, *Commentary on the Almagest (Greek)*' (update: 03.01.2023), *Ptolemaeus Arabus et Latinus. Works*, URL = http://ptolemaeus.badw.de/work/83.

Juste, David [85]. 'Theon of Alexandria, ⟨*Commentum in Almagesti*⟩ (tr. Giovanni Battista Teofilo)' (update: 02.12.2022), *Ptolemaeus Arabus et Latinus. Works*, URL = http://ptolemaeus.badw.de/work/85.

Juste, David [164]. 'Theon of Alexandria, ⟨*Commentum in Almagesti*⟩ (tr. Jerónimo Muñoz)' (update: 02.12.2022), *Ptolemaeus Arabus et Latinus. Works*, URL = http://ptolemaeus.badw.de/work/164.

Juste, David [165]. 'Theon of Alexandria, ⟨*Commentum in Almagesti*⟩ (tr. late 16th c.)' (update: 02.12.2022), *Ptolemaeus Arabus et Latinus. Works*, URL = http://ptolemaeus.badw.de/work/165.

Index